POCKET GUIDE TO
RESIDENTIAL ELECTRICAL
INSTALLATIONS

2011 Edition

CHARLES R. MILLER

WITHDRAWN

National Fire Protection Association®
Quincy, Massachusetts

Project Manager: Ann Coughlin
Project Editor: Irene Herlihy
Editorial-Production Services: Omegatype Typography, Inc.
Interior Design: Omegatype Typography, Inc.
Composition: Omegatype Typography, Inc.
Cover Design: Cameron, Inc.
Manufacturing Manager: Ellen Glisker
Printer: Dickinson Press, Inc.

The following are registered trademarks of the National Fire
Protection Association:

National Fire Protection Association®
NFPA®

National Electrical Code®
NEC®

NFPA No.: PGNECRES11
ISBN: 978-0-87765-965-5

Library of Congress Control Number: 2010931041

Printed in the United States of America

10 11 12 13   5 4 3 2 1

IMPORTANT NOTICES AND DISCLAIMERS OF LIABILITY

This *Pocket Guide* is a compilation of extracts from the 2011 edition of the *National Electrical Code®* (*NEC®*). These extracts have been selected and arranged by the Author of the *Pocket Guide*, who has also supplied introductory material generally located at the beginning of each chapter.

The *NEC*, like all NFPA® codes and standards, is a document that is developed through a consensus standards development process approved by the American National Standards Institute. This process brings together volunteers representing varied viewpoints and interests to achieve consensus on fire, electrical, and other safety issues. While the NFPA administers the process and establishes rules to promote fairness in the development of consensus, it does not independently test, evaluate, or verify the accuracy of any information or the soundness of any judgments contained in its codes and standards.

The NFPA and the Author disclaim liability for any personal injury, property or other damages of any nature whatsoever, whether special, indirect, consequential, or compensatory, directly or indirectly resulting from the publication, use of, or reliance on the *NEC* or this *Pocket Guide*. The NFPA and the Author also make no guaranty or warranty as to the accuracy or completeness of any information published herein.

In issuing and making the *NEC* and this *Pocket Guide* available, the NFPA and the Author are not undertaking to render professional or other services for or on behalf of any person or entity. Nor is the NFPA or the Author undertaking to perform any duty owed by any person or entity to someone else. Anyone using these materials should rely on his or her own independent judgment or, as appropriate, seek the advice of a competent professional in determining the exercise of reasonable care in any given circumstances.

This *Pocket Guide* is not intended to be a substitute for the *NEC* itself, which should always be consulted to ensure full *Code* compliance.

Additional Important Notices and Legal Disclaimers concerning the use of the *NEC* can be viewed at www.nfpa.org/disclaimers.

NOTICE CONCERNING CODE INTERPRETATIONS

The introductory materials and the selection and arrangement of the *NEC* extracts contained in this *Pocket Guide* reflect the personal opinions and judgments of the Author and do not necessarily represent the official position of the NFPA (which can only be obtained through Formal Interpretations processed in accordance with NFPA rules).

CONTENTS/TABS

CONTENTS/TABS

CHAPTER/ARTICLE CROSS-REFERENCE TABLE

Note: *NEC* article number refers to selected requirements from that article.

Chapter No.	*NEC* Article No.
1	90
2	100
3	110
4	590
5	220
6	230
7	215
8	408, 404
9	240
10	200
11	250
12	314
13	334, 330, 338, 340
14	358, 344, 352, 362, 348, 350, 356
15	300, 310
16	210
17	210, 406
18	404
19	210, 410, 411
20	422
21	424
22	440
23	800
24	225, 250
25	680
26	Chapter 9 Tables, Annex C

ARTICLE/CHAPTER CROSS-REFERENCE TABLE

Note: *NEC* article number refers to selected requirements from that article

NEC Article No.	Chapter No.
90	1
100	2
110	3
200	10
210	16, 17, 19
215	7
220	5
225	24
230	6
240	9
250	11, 24
300	15
310	15
314	12
330	13
334	13
338	13
340	13
344	14
348	14
350	14
352	14
356	14
358	14
362	14
404	8, 18
406	17
408	8
410	19
411	19
422	20
424	21
440	22
590	4
680	25
800	23
Annex C	26
Chapter 9 Tables	26

INTRODUCTION

The *Pocket Guide to Residential Electrical Installations* has been developed to provide a portable and practical comprehensive field reference source for electricians, contractors, inspectors, and others responsible for applying the rules of the *National Electrical Code®*. Together with its companion volume, *Pocket Guide to Commercial and Industrial Electrical Installations*, the most widely applied general requirements of the *NEC®* are compiled in these two extremely useful on-the-job, on-the-go documents.

The compiled requirements follow an order similar to the progression of an electrical system installation on the construction site, rather than the normal sequence of requirements in the *NEC*. Introductory chapters are followed by the requirements pertaining to temporary electrical installations. The *Pocket Guide* then continues with branch circuit, feeder, and service load calculation requirements from Article 220 and then proceeds into other general requirements extracted from Chapters 2, 3, and 4 of the *NEC*.

Due to its compact size, the *Pocket Guide* does not contain all of the *Code* requirements and is a complement to, not a replacement of, the 2011 *NEC*. When paired with the entire volume of electrical safety requirements contained in the 2011 *Code*, the *Pocket Guide* is an effective tool that everyone concerned with safe electrical installations will want to have. In-depth explanations of *NEC* requirements are available in NFPA's *National Electrical Code Handbook*, the only *NEC* handbook with the entire text of the *NEC*, plus useful explanation and commentary developed by the NFPA staff. Users of the *Pocket Guide to Residential Electrical Installations* must also understand that some adopting jurisdictions enact requirements to address unique local environmental, geographical, or other considerations. Always consult with the local authority having jurisdiction for information on locally enacted requirements.

The highly specialized requirements contained in Chapters 5 through 8 of the *Code* are not included in this compilation, as this material is outside of the intended purpose of this document. The requirements included in the *Pocket Guide* are those that apply to virtually all residential electrical installations and are based on a practical working knowledge of how a residential electrical construction project evolves from beginning to completion. As indicated above, however, the *Pocket Guide* is not intended to be a substitute for the *NEC* itself, which should always be consulted to ensure full *Code* compliance.

It is hoped that this 2011 edition of the *Pocket Guide to Residential Electrical Installations* proves to be one of the most valuable tools that you use on a daily basis in performing safe, *NEC*-compliant electrical installations.

POCKET GUIDE TO
RESIDENTIAL ELECTRICAL
INSTALLATIONS

CHAPTER 1

INTRODUCTION TO THE
NATIONAL ELECTRICAL CODE

INTRODUCTION

This chapter contains provisions from Article 90—Introduction. Article 90 covers basic provisions essential for understanding and applying the *National Electrical Code®*. Besides providing the purpose of the *Code,* Article 90 specifies the document scope, describes how the *Code* is arranged, provides information on enforcement, and contains other provisions that are essential to the proper use and application of the *Code.* The first sentence states that "the purpose of this *Code* is the practical safeguarding of persons and property from hazards arising from the use of electricity." Safety, for both property and people, is the main objective for the provisions found in the *Code.* Compliance with these provisions and proper maintenance will result in an installation essentially free from electrical hazards but not necessarily efficient, convenient, or adequate for good service or future expansion of electrical use. Because of the limited space within this pocket guide, only part of Article 90 has been included. Reference the 2011 *National Electrical Code* for all of the requirements.

   Although state and local jurisdictions generally adopt the *Code* exactly as written, they may also amend the *Code* by incorporating local requirements. Therefore, obtain a copy of any additional rules and regulations for your area.

ARTICLE 90
Introduction

90.1 Purpose.

(A) Practical Safeguarding. The purpose of this *Code* is the practical safeguarding of persons and property from hazards arising from the use of electricity.

(B) Adequacy. This *Code* contains provisions that are considered necessary for safety. Compliance therewith and proper maintenance results in an installation that is essentially free from hazard but not necessarily efficient, convenient, or adequate for good service or future expansion of electrical use.

90.4 Enforcement. This *Code* is intended to be suitable for mandatory application by governmental bodies that exercise legal jurisdiction over electrical installations, including signaling and communications systems, and for use by insurance inspectors. The authority having jurisdiction for enforcement of the *Code* has the responsibility for making interpretations of the rules, for deciding on the approval of equipment and materials, and for granting the special permission contemplated in a number of the rules.

By special permission, the authority having jurisdiction may waive specific requirements in this *Code* or permit alternative methods where it is assured that equivalent objectives can be achieved by establishing and maintaining effective safety.

This *Code* may require new products, constructions, or materials that may not yet be available at the time the *Code* is adopted. In such event, the authority having jurisdiction may permit the use of the products, constructions, or materials that comply with the most recent previous edition of this *Code* adopted by the jurisdiction.

90.5 Mandatory Rules, Permissive Rules, and Explanatory Material.

(A) Mandatory Rules. Mandatory rules of this *Code* are those that identify actions that are specifically required or prohibited and are characterized by the use of the terms shall or shall not.

(B) Permissive Rules. Permissive rules of this *Code* are those that identify actions that are allowed but not required, are normally used to describe options or alternative methods, and are characterized by the use of the terms shall be permitted or shall not be required.

(C) Explanatory Material. Explanatory material, such as references to other standards, references to related sections of this *Code*, or information related to a *Code* rule, is included in this *Code* in the form of informational notes. Such notes are informational only and are not enforceable as requirements of this *Code*.

Brackets containing section references to another NFPA document are for informational purposes only and are provided as a guide to indicate the source of the extracted text. These bracketed references immediately follow the extracted text.

(D) Informative Annexes. Nonmandatory information relative to the use of the *NEC* is provided in informative annexes. Informative annexes are not part of the enforceable requirements of the *NEC*, but are included for information purposes only.

90.8 Wiring Planning.

(A) Future Expansion and Convenience. Plans and specifications that provide ample space in raceways, spare raceways, and additional spaces allow for future increases in electric power and communications circuits. Distribution centers located in readily accessible locations provide convenience and safety of operation.

(B) Number of Circuits in Enclosures. It is elsewhere provided in this *Code* that the number of wires and circuits confined in a single enclosure be varyingly restricted. Limiting the number of circuits in a single enclosure minimizes the effects from a short circuit or ground fault in one circuit.

CHAPTER 2

DEFINITIONS APPLICABLE
TO DWELLING UNITS

INTRODUCTION

The definition of a word or phrase used in the *NEC* may differ from that found in a dictionary. In fact, some of these terms are not defined in a standard dictionary. Understanding the *Code* without a thorough understanding of the terms is difficult. Therefore, having an understanding of Article 100 is an important step toward proper application of the *Code*.

As stated in the scope of Article 100, only definitions essential to the proper application of the *Code* are included. Since commonly defined general terms can be found in most dictionaries, they are not included. Also not included are commonly defined technical terms. These terms can be found in a dictionary of electrical and electronic terms.

Since a number of terms in Article 100 do not apply to dwelling units, they are not included in this book. Likewise, because of this guide's compact design, other terms and informational notes are also omitted. Reference the 2011 *National Electrical Code* for a complete listing.

Article 100 contains terms used in two or more articles. Other terms that pertain to one article may be defined in that particular article. For example, Article 680—Swimming Pools, Fountains, and Similar Installations includes terms that are not used anywhere else in the *Code*. Therefore, those terms are defined in Article 680 rather than in Article 100.

ARTICLE 100
Definitions

Scope. This article contains only those definitions essential to the proper application of this *Code*. It is not intended to include commonly defined general terms or commonly defined technical terms from related codes and standards. In general, only those terms that are used in two or more articles are defined in Article 100. Other definitions are included in the article in which they are used but may be referenced in Article 100.

Part I of this article contains definitions intended to apply wherever the terms are used throughout this *Code*. Part II contains definitions applicable only to the parts of articles specifically covering installations and equipment operating at over 600 volts, nominal.

I. General

Accessible (as applied to equipment). Admitting close approach; not guarded by locked doors, elevation, or other effective means.

Accessible (as applied to wiring methods). Capable of being removed or exposed without damaging the building structure or finish or not permanently closed in by the structure or finish of the building.

Accessible, Readily (Readily Accessible). Capable of being reached quickly for operation, renewal, or inspections without requiring those to whom ready access is requisite to climb over or remove obstacles or to resort to portable ladders, and so forth.

Ampacity. The maximum current, in amperes, that a conductor can carry continuously under the conditions of use without exceeding its temperature rating.

Appliance. Utilization equipment, generally other than industrial, that is normally built in standardized sizes or types and is installed or connected as a unit to perform one or more functions such as clothes washing, air conditioning, food mixing, deep frying, and so forth.

Approved. Acceptable to the authority having jurisdiction.

Arc-Fault Circuit Interrupter (AFCI). A device intended to provide protection from the effects of arc faults by recognizing characteristics unique to arcing and by functioning to de-energize the circuit when an arc fault is detected.

Attachment Plug (Plug Cap) (Plug). A device that, by insertion in a receptacle, establishes a connection between the conductors of the attached flexible cord and the conductors connected permanently to the receptacle.

Authority Having Jurisdiction (AHJ). An organization, office, or individual responsible for enforcing the requirements of a code or standard, or for approving equipment, materials, an installation, or a procedure.

> Informational Note: The phrase "authority having jurisdiction," or its acronym AHJ, is used in NFPA documents in a broad manner, since jurisdictions and approval agencies vary, as do their responsibilities. Where public safety is primary, the authority having jurisdiction may be a federal, state, local, or other regional department or individual such as a fire chief; fire marshal; chief of a fire prevention bureau, labor department, or health department; building official; electrical inspector; or others having statutory authority. For insurance purposes, an insurance inspection department, rating bureau, or other insurance company representative may be the authority having jurisdiction. In many circumstances, the property owner or his or her designated agent assumes the role of the authority having jurisdiction; at government installations, the commanding officer or departmental official may be the authority having jurisdiction.

Automatic. Performing a function without the necessity of human intervention.

Bathroom. An area including a basin with one or more of the following: a toilet, a urinal, a tub, a shower, a bidet, or similar plumbing fixtures.

Bonded (Bonding). Connected to establish electrical continuity and conductivity.

Bonding Conductor or Jumper. A reliable conductor to ensure the required electrical conductivity between metal parts required to be electrically connected.

Bonding Jumper, Equipment. The connection between two or more portions of the equipment grounding conductor.

Bonding Jumper, Main. The connection between the grounded circuit conductor and the equipment grounding conductor at the service.

Branch Circuit. The circuit conductors between the final overcurrent device protecting the circuit and the outlet(s).

Branch Circuit, Appliance. A branch circuit that supplies energy to one or more outlets to which appliances are to be connected and that has no permanently connected luminaires that are not a part of an appliance.

Branch Circuit, General-Purpose. A branch circuit that supplies two or more receptacles or outlets for lighting and appliances.

Branch Circuit, Individual. A branch circuit that supplies only one utilization equipment.

Branch Circuit, Multiwire. A branch circuit that consists of two or more ungrounded conductors that have a voltage between them, and a grounded conductor that has equal voltage between it and each ungrounded conductor of the circuit and that is connected to the neutral or grounded conductor of the system.

Building. A structure that stands alone or that is cut off from adjoining structures by fire walls with all openings therein protected by approved fire doors.

Cabinet. An enclosure that is designed for either surface mounting or flush mounting and is provided with a frame, mat, or trim in which a swinging door or doors are or can be hung.

Circuit Breaker. A device designed to open and close a circuit by nonautomatic means and to open the circuit automatically on a predetermined overcurrent without damage to itself when properly applied within its rating.

Clothes Closet. A non-habitable room or space intended primarily for storage of garments and apparel.

Communications Equipment. The electronic equipment that performs the telecommunications operations for the transmission of audio, video, and data, and includes power equipment (e.g., dc converters, inverters, and batteries) and technical support equipment (e.g., computers).

Concealed. Rendered inaccessible by the structure or finish of the building. Wires in concealed raceways are considered concealed, even though they may become accessible by withdrawing them.

Conductor, Bare. A conductor having no covering or electrical insulation whatsoever.

Conductor, Covered. A conductor encased within material of composition or thickness that is not recognized by this *Code* as electrical insulation.

Conductor, Insulated. A conductor encased within material of composition and thickness that is recognized by this *Code* as electrical insulation.

Conduit Body. A separate portion of a conduit or tubing system that provides access through a removable cover(s) to the interior of the system at a junction of two or more sections of the system or at a terminal point of the system.
    Boxes such as FS and FD or larger cast or sheet metal boxes are not classified as conduit bodies.

Connector, Pressure (Solderless). A device that establishes a connection between two or more conductors or between one or more conductors and a terminal by means of mechanical pressure and without the use of solder.

Continuous Load. A load where the maximum current is expected to continue for 3 hours or more.

Cooking Unit, Counter-Mounted. A cooking appliance designed for mounting in or on a counter and consisting of one or more heating elements, internal wiring, and built-in or mountable controls.

Copper-Clad Aluminum Conductors. Conductors drawn from a copper-clad aluminum rod with the copper metallurgically bonded to an aluminum core. The copper forms a minimum of 10 percent of the cross-sectional area of a solid conductor or each strand of a stranded conductor.

Cutout Box. An enclosure designed for surface mounting that has swinging doors or covers secured directly to and telescoping with the walls of the box proper.

Dead Front. Without live parts exposed to a person on the operating side of the equipment.

Demand Factor. The ratio of the maximum demand of a system, or part of a system, to the total connected load of a system or the part of the system under consideration.

Device. A unit of an electrical system that carries or controls electric energy as its principal function.

Disconnecting Means. A device, or group of devices, or other means by which the conductors of a circuit can be disconnected from their source of supply.

Dwelling, One-Family. A building that consists solely of one dwelling unit.

Dwelling, Two-Family. A building that consists solely of two dwelling units.

Dwelling, Multifamily. A building that contains three or more dwelling units.

Dwelling Unit. A single unit, providing complete and independent living facilities for one or more persons, including permanent provisions for living, sleeping, cooking, and sanitation.

Enclosed. Surrounded by a case, housing, fence, or wall(s) that prevents persons from accidentally contacting energized parts.

Enclosure. The case or housing of apparatus, or the fence or walls surrounding an installation to prevent personnel from accidentally contacting energized parts or to protect the equipment from physical damage.

Energized. Electrically connected to, or is, a source of voltage.

Equipment. A general term, including fittings, devices, appliances, luminaires, apparatus, machinery, and the like used as a part of, or in connection with, an electrical installation.

Exposed (as applied to live parts). Capable of being inadvertently touched or approached nearer than a safe distance by a person. It is applied to parts that are not suitably guarded, isolated, or insulated.

Exposed (as applied to wiring methods). On or attached to the surface or behind panels designed to allow access.

Feeder. All circuit conductors between the service equipment, the source of a separately derived system, or other power supply source and the final branch-circuit overcurrent device.

Festoon Lighting. A string of outdoor lights that is suspended between two points.

Fitting. An accessory such as a locknut, bushing, or other part of a wiring system that is intended primarily to perform a mechanical rather than an electrical function.

Garage. A building or portion of a building in which one or more self-propelled vehicles can be kept for use, sale, storage, rental, repair, exhibition, or demonstration purposes.

Grounded (Grounding). Connected (connecting) to ground or to a conductive body that extends the ground connection.

Grounded, Solidly. Connected to ground without inserting any resistor or impedance device.

Grounded Conductor. A system or circuit conductor that is intentionally grounded.

Ground-Fault Circuit Interrupter (GFCI). A device intended for the protection of personnel that functions to de-energize a circuit or portion thereof within an established period of time when a current to ground exceeds the values established for a Class A device.

Grounding Conductor, Equipment (EGC). The conductive path(s) installed to connect normally non–current-carrying metal parts of equipment together and to the system grounded conductor or to the grounding electrode conductor, or both.

Grounding Electrode. A conducting object through which a direct connection to earth is established.

Grounding Electrode Conductor. A conductor used to connect the system grounded conductor or the equipment to a grounding electrode or to a point on the grounding electrode system.

Guarded. Covered, shielded, fenced, enclosed, or otherwise protected by means of suitable covers, casings, barriers, rails, screens, mats, or platforms to remove the likelihood of approach or contact by persons or objects to a point of danger.

Guest Room. An accommodation combining living, sleeping, sanitary, and storage facilities within a compartment.

Handhole Enclosure. An enclosure for use in underground systems, provided with an open or closed bottom, and sized to allow personnel to reach into, but not enter, for the purpose of installing, operating, or maintaining equipment or wiring or both.

Identified (as applied to equipment). Recognizable as suitable for the specific purpose, function, use, environment, application, and so forth, where described in a particular *Code* requirement.

In Sight From (Within Sight From, Within Sight). Where this *Code* specifies that one equipment shall be "in sight from," "within sight from," or "within sight of," and so forth, another equipment, the specified equipment is to be visible and not more than 15 m (50 ft) distant from the other.

Intersystem Bonding Termination. A device that provides a means for connecting bonding conductors for communications systems to the grounding electrode system.

Kitchen. An area with a sink and permanent provisions for food preparation and cooking.

Labeled. Equipment or materials to which has been attached a label, symbol, or other identifying mark of an organization that is acceptable to the authority having jurisdiction and concerned with product evaluation, that maintains periodic inspection of production of labeled equipment or materials, and by whose labeling the manufacturer indicates compliance with appropriate standards or performance in a specified manner.

Lighting Outlet. An outlet intended for the direct connection of a lampholder or luminaire.

Listed. Equipment, materials, or services included in a list published by an organization that is acceptable to the authority having jurisdiction and concerned with evaluation of products or services, that maintains periodic inspection of production of listed equipment or materials or periodic evaluation of services, and whose listing states that either the equipment, material, or service meets appropriate designated standards or has been tested and found suitable for a specified purpose.

Location, Damp. Locations protected from weather and not subject to saturation with water or other liquids but subject to moderate degrees of moisture. Examples of such locations include partially protected locations under canopies, marquees, roofed open porches, and like locations, and interior locations subject to moderate degrees of moisture, such as some basements, some barns, and some cold-storage warehouses.

Location, Dry. A location not normally subject to dampness or wetness. A location classified as dry may be temporarily subject to dampness or wetness, as in the case of a building under construction.

Location, Wet. Installations underground or in concrete slabs or masonry in direct contact with the earth; in locations subject to saturation with water or other liquids, such as vehicle washing areas; and in unprotected locations exposed to weather.

Luminaire. A complete lighting unit consisting of a light source such as a lamp or lamps, together with the parts designed to position the light source and connect it to the power supply. It may also include parts to protect the light source or the ballast or to distribute the light. A lampholder itself is not a luminaire.

Multioutlet Assembly. A type of surface, flush, or free-standing raceway designed to hold conductors and receptacles, assembled in the field or at the factory.

Neutral Conductor. The conductor connected to the neutral point of a system that is intended to carry current under normal conditions.

Neutral Point. The common point on a wye-connection in a polyphase system or midpoint on a single-phase, 3-wire system, or midpoint of a single-phase portion of a 3-phase delta system, or a midpoint of a 3-wire, direct-current system.

Outlet. A point on the wiring system at which current is taken to supply utilization equipment.

Outline Lighting. An arrangement of incandescent lamps, electric-discharge lighting, or other electrically powered light sources to outline or call attention to certain features such as the shape of a building or the decoration of a window.

Overcurrent. Any current in excess of the rated current of equipment or the ampacity of a conductor. It may result from overload, short circuit, or ground fault.

Overcurrent Protective Device, Branch-Circuit. A device capable of providing protection for service, feeder, and branch circuits and equipment over the full range of overcurrents between its rated current and its interrupting rating. Branch-circuit overcurrent protective devices are provided with interrupting ratings appropriate for the intended use but no less than 5000 amperes.

Overcurrent Protective Device, Supplementary. A device intended to provide limited overcurrent protection for specific applications and utilization equipment such as luminaires and appliances. This limited protection is in addition to the protection provided in the required branch circuit by the branch-circuit overcurrent protective device.

Overload. Operation of equipment in excess of normal, full-load rating, or of a conductor in excess of rated ampacity that, when it persists for a sufficient length of time, would cause damage or dangerous overheating. A fault, such as a short circuit or ground fault, is not an overload.

Panelboard. A single panel or group of panel units designed for assembly in the form of a single panel, including buses and automatic overcurrent devices, and equipped with or without switches for the control of light, heat, or power circuits; designed to be placed in a cabinet or cutout box placed in or against a wall, partition, or other support; and accessible only from the front.

Premises Wiring (System). Interior and exterior wiring, including power, lighting, control, and signal circuit wiring together with all their associated hardware, fittings, and wiring devices, both permanently and temporarily installed. This includes (a) wiring from the service point or power source to the outlets or (b) wiring from and including the power source to the outlets where there is no service point.

Such wiring does not include wiring internal to appliances, luminaires, motors, controllers, motor control centers, and similar equipment.

Raceway. An enclosed channel of metal or nonmetallic materials designed expressly for holding wires, cables, or busbars, with additional functions as permitted in this *Code*. Raceways include, but are not limited to, rigid metal conduit, rigid nonmetallic conduit, intermediate metal conduit, liquidtight flexible conduit, flexible metallic tubing, flexible metal conduit, electrical nonmetallic tubing, electrical metallic tubing, underfloor raceways, cellular concrete floor raceways, cellular metal floor raceways, surface raceways, wireways, and busways.

Rainproof. Constructed, protected, or treated so as to prevent rain from interfering with the successful operation of the apparatus under specified test conditions.

Raintight. Constructed or protected so that exposure to a beating rain will not result in the entrance of water under specified test conditions.

Receptacle. A receptacle is a contact device installed at the outlet for the connection of an attachment plug. A single receptacle is a single contact device with no other contact device on the same yoke. A multiple receptacle is two or more contact devices on the same yoke.

Receptacle Outlet. An outlet where one or more receptacles are installed.

Remote-Control Circuit. Any electrical circuit that controls any other circuit through a relay or an equivalent device.

Separately Derived System. A premises wiring system whose power is derived from a source of electric energy or equipment other than a service. Such systems have no direct connection from circuit conductors of one system to circuit conductors of another system, other than connections through the earth, metal enclosures, metallic raceways, or equipment grounding conductors.

Service. The conductors and equipment for delivering electric energy from the serving utility to the wiring system of the premises served.

Service Cable. Service conductors made up in the form of a cable.

Service Conductors. The conductors from the service point to the service disconnecting means.

Service Conductors, Overhead. The overhead conductors between the service point and the first point of connection to the service-entrance conductors at the building or other structure.

Service Conductors, Underground. The underground conductors between the service point and the first point of connection to the service-entrance conductors in a terminal box, meter, or other enclosure, inside or outside the building wall.

Service Drop. The overhead conductors between the utility electric supply system and the service point.

Service-Entrance Conductors, Overhead System. The service conductors between the terminals of the service equipment and a point usually outside the building, clear of building walls, where joined by tap or splice to the service drop or overhead service conductors.

Service-Entrance Conductors, Underground System. The service conductors between the terminals of the service equipment and the point of connection to the service lateral or underground service conductors.

Service Equipment. The necessary equipment, usually consisting of a circuit breaker(s) or switch(es) and fuse(s) and their accessories, connected to the load end of service conductors to a building or other structure, or an otherwise designated area, and intended to constitute the main control and cutoff of the supply.

Service Lateral. The underground conductors between the utility electric supply system and the service point.

Service Point. The point of connection between the facilities of the serving utility and the premises wiring.

> Informational Note: The service point can be described as the point of demarcation between where the serving utility ends and the premises wiring begins. The serving utility generally specifies the location of the service point based on the conditions of service.

Solar Photovoltaic System. The total components and subsystems that, in combination, convert solar energy into electric energy suitable for connection to a utilization load.

Special Permission. The written consent of the authority having jurisdiction.

Structure. That which is built or constructed.

Surge Arrester. A protective device for limiting surge voltages by discharging or bypassing surge current; it also prevents continued flow of follow current while remaining capable of repeating these functions.

Surge-Protective Device (SPD). A protective device for limiting transient voltages by diverting or limiting surge current; it also prevents continued flow of follow current while remaining capable of repeating these functions and is designated as follows:

Type 1: Permanently connected SPDs intended for installation between the secondary of the service transformer and the line side of the service disconnect overcurrent device.

Type 2: Permanently connected SPDs intended for installation on the load side of the service disconnect overcurrent device, including SPDs located at the branch panel.

Type 3: Point of utilization SPDs.

Type 4: Component SPDs, including discrete components, as well as assemblies.

Switch, General-Use. A switch intended for use in general distribution and branch circuits. It is rated in amperes, and it is capable of interrupting its rated current at its rated voltage.

Switch, General-Use Snap. A form of general-use switch constructed so that it can be installed in device boxes or on box covers, or otherwise used in conjunction with wiring systems recognized by this *Code*.

Switch, Transfer. An automatic or nonautomatic device for transferring one or more load conductor connections from one power source to another.

Ungrounded. Not connected to ground or to a conductive body that extends the ground connection.

Utilization Equipment. Equipment that utilizes electric energy for electronic, electromechanical, chemical, heating, lighting, or similar purposes.

Voltage (of a circuit). The greatest root-mean-square (rms) (effective) difference of potential between any two conductors of the circuit concerned.

Voltage, Nominal. A nominal value assigned to a circuit or system for the purpose of conveniently designating its voltage class (e.g., 120/240 volts, 480Y/277 volts, 600 volts). The actual voltage at which a circuit operates can vary from the nominal within a range that permits satisfactory operation of equipment.

Voltage to Ground. For grounded circuits, the voltage between the given conductor and that point or conductor of the circuit that is grounded; for ungrounded circuits, the greatest voltage between the given conductor and any other conductor of the circuit.

Watertight. Constructed so that moisture will not enter the enclosure under specified test conditions.

Weatherproof. Constructed or protected so that exposure to the weather will not interfere with successful operation.

> Informational Note: Rainproof, raintight, or watertight equipment can fulfill the requirements for weatherproof where varying weather conditions other than wetness, such as snow, ice, dust, or temperature extremes, are not a factor.

CHAPTER 3

GENERAL INSTALLATION REQUIREMENTS

INTRODUCTION

This chapter contains provisions from Article 110—Requirements for Electrical Installations. These provisions are essential to the installation of conductors and equipment. Article 110 provides fundamental safety requirements that apply to all types of electrical installations. While certain provisions apply to all aspects of electrical installations, others pertain to specific areas. For example, conductors and equipment required (or permitted) by the *Code* must be approved [110.2]. While this requirement applies to the installation of all conductors and equipment, other provisions are more specific. For example, the requirements of 110.26(F) apply only to switchboards, panelboards, distribution boards, and motor control centers.

ARTICLE 110
Requirements for Electrical Installations

I. GENERAL

110.1 Scope. This article covers general requirements for the examination and approval, installation and use, access to and spaces about electrical conductors and equipment; enclosures intended for personnel entry; and tunnel installations.

110.2 Approval. The conductors and equipment required or permitted by this *Code* shall be acceptable only if approved.

110.3 Examination, Identification, Installation, and Use of Equipment.

(A) Examination. In judging equipment, considerations such as the following shall be evaluated:

(1) Suitability for installation and use in conformity with the provisions of this *Code*
(2) Mechanical strength and durability, including, for parts designed to enclose and protect other equipment, the adequacy of the protection thus provided
(3) Wire-bending and connection space
(4) Electrical insulation
(5) Heating effects under normal conditions of use and also under abnormal conditions likely to arise in service
(6) Arcing effects
(7) Classification by type, size, voltage, current capacity, and specific use
(8) Other factors that contribute to the practical safeguarding of persons using or likely to come in contact with the equipment

(B) Installation and Use. Listed or labeled equipment shall be installed and used in accordance with any instructions included in the listing or labeling.

110.5 Conductors. Conductors normally used to carry current shall be of copper unless otherwise provided in this *Code*. Where the conductor material is not specified, the material and the sizes given in this *Code* shall apply to copper conductors. Where other materials are used, the size shall be changed accordingly.

110.6 Sizes. Conductor sizes are expressed in American Wire Gage (AWG) or in circular mils.

110.7 Wiring Integrity. Completed wiring installations shall be free from short circuits, ground faults, or any connections to ground other than as required or permitted elsewhere in this *Code.*

110.11 Deteriorating Agents. Unless identified for use in the operating environment, no conductors or equipment shall be located in damp or wet locations; where exposed to gases, fumes, vapors, liquids, or other agents that have a deteriorating effect on the conductors or equipment; or where exposed to excessive temperatures.

Equipment not identified for outdoor use and equipment identified only for indoor use, such as "dry locations," "indoor use only," "damp locations," or enclosure Types 1, 2, 5, 12, 12K, and/or 13, shall be protected against damage from the weather during construction.

110.12 Mechanical Execution of Work. Electrical equipment shall be installed in a neat and workmanlike manner.

(A) Unused Openings. Unused openings, other than those intended for the operation of equipment, those intended for mounting purposes, or those permitted as part of the design for listed equipment, shall be closed to afford protection substantially equivalent to the wall of the equipment. Where metallic plugs or plates are used with nonmetallic enclosures, they shall be recessed at least 6 mm (1/4 in.) from the outer surface of the enclosure.

(B) Integrity of Electrical Equipment and Connections. Internal parts of electrical equipment, including busbars, wiring terminals, insulators, and other surfaces, shall not be damaged or contaminated by foreign materials such as paint, plaster, cleaners, abrasives, or corrosive residues. There shall be no damaged parts that may adversely affect safe operation or mechanical strength of the equipment such as parts that are broken; bent; cut; or deteriorated by corrosion, chemical action, or overheating.

110.13 Mounting and Cooling of Equipment.

(A) Mounting. Electrical equipment shall be firmly secured to the surface on which it is mounted. Wooden plugs driven into holes in masonry, concrete, plaster, or similar materials shall not be used.

110.14 Electrical Connections. Because of different characteristics of dissimilar metals, devices such as pressure terminal or pressure splicing connectors and soldering lugs shall be identified for the material of the conductor and shall be properly installed and used. Conductors of dissimilar metals shall not be intermixed in a terminal or splicing connector where physical contact occurs between dissimilar conductors (such as copper and aluminum, copper and copper-clad aluminum, or aluminum and copper-clad aluminum), unless the device is identified for the purpose and conditions of use. Materials such as solder, fluxes, inhibitors, and compounds, where employed, shall be suitable for the use and shall be of a type that will not adversely affect the conductors, installation, or equipment.

(A) Terminals. Connection of conductors to terminal parts shall ensure a thoroughly good connection without damaging the conductors and shall be made by means of pressure connectors (including set-screw type), solder lugs, or splices to flexible leads. Connection by means of wire-binding screws or studs and nuts that have upturned lugs or the equivalent shall be permitted for 10 AWG or smaller conductors.

Terminals for more than one conductor and terminals used to connect aluminum shall be so identified.

(B) Splices. Conductors shall be spliced or joined with splicing devices identified for the use or by brazing, welding, or soldering with a fusible metal or alloy. Soldered splices shall first be spliced or joined so as to be mechanically and electrically secure without solder and then be soldered. All splices and joints and the free ends of conductors shall be covered with an insulation equivalent to that of the conductors or with an insulating device identified for the purpose.

Wire connectors or splicing means installed on conductors for direct burial shall be listed for such use.

(C) Temperature Limitations. The temperature rating associated with the ampacity of a conductor shall be selected and coordinated so as not to exceed the lowest temperature rating of any connected termination, conductor, or device. Conductors with temperature ratings higher than specified for terminations shall be permitted to be used for ampacity adjustment, correction, or both.

(1) Equipment Provisions. The determination of termination provisions of equipment shall be based on 110.14(C)(1)(a) or (C)(1)(b). Unless the equipment is listed and marked otherwise, conductor ampacities used in determining equipment termination provisions shall be based on Table 310.15(B)(16) as appropriately modified by 310.15(B)(6).

(a) Termination provisions of equipment for circuits rated 100 amperes or less, or marked for 14 AWG through 1 AWG conductors, shall be used only for one of the following:

(1) Conductors rated 60°C (140°F).
(2) Conductors with higher temperature ratings, provided the ampacity of such conductors is determined based on the 60°C (140°F) ampacity of the conductor size used.
(3) Conductors with higher temperature ratings if the equipment is listed and identified for use with such conductors.
(4) For motors marked with design letters B, C, or D, conductors having an insulation rating of 75°C (167°F) or higher shall be permitted to be used, provided the ampacity of such conductors does not exceed the 75°C (167°F) ampacity.

(b) Termination provisions of equipment for circuits rated over 100 amperes, or marked for conductors larger than 1 AWG, shall be used only for one of the following:

(1) Conductors rated 75°C (167°F)
(2) Conductors with higher temperature ratings, provided the ampacity of such conductors does not exceed the 75°C (167°F) ampacity of the conductor size used, or up to their ampacity if the equipment is listed and identified for use with such conductors

110.22 Identification of Disconnecting Means.

(A) General. Each disconnecting means shall be legibly marked to indicate its purpose unless located and arranged so the purpose is evident. The marking shall be of sufficient durability to withstand the environment involved.

II. 600 VOLTS, NOMINAL, OR LESS

110.26 Spaces About Electrical Equipment. Access and working space shall be provided and maintained about all electrical equipment to permit ready and safe operation and maintenance of such equipment.

(A) Working Space. Working space for equipment operating at 600 volts, nominal, or less to ground and likely to require examination, adjustment, servicing, or maintenance while energized shall comply with the dimensions of 110.26(A)(1), (A)(2), and (A)(3) or as required or permitted elsewhere in this *Code.*

(1) Depth of Working Space. The depth of the working space in the direction of live parts shall not be less than that specified in Table 110.26(A)(1) unless the requirements of 110.26(A)(1)(a), (A)(1)(b), or (A)(1)(c) are met. Distances shall be measured from the exposed live parts or from the enclosure or opening if the live parts are enclosed.

Table 110.26(A)(1) Working Spaces

Nominal Voltage to Ground	Minimum Clear Distance		
	Condition 1	Condition 2	Condition 3
0–150	914 mm (3 ft)	914 mm (3 ft)	914 mm (3 ft)
151–600	914 mm (3 ft)	1.07 m (3 ft 6 in.)	1.22 m (4 ft)

Note: Where the conditions are as follows:

Condition 1—Exposed live parts on one side of the working space and no live or grounded parts on the other side of the working space, or exposed live parts on both sides of the working space that are effectively guarded by insulating materials.

Condition 2—Exposed live parts on one side of the working space and grounded parts on the other side of the working space. Concrete, brick, or tile walls shall be considered as grounded.

Condition 3—Exposed live parts on both sides of the working space.

(2) Width of Working Space The width of the working space in front of the electrical equipment shall be the width of the equipment or 762 mm (30 in.), whichever is greater. In all cases, the work space shall permit at least a 90 degree opening of equipment doors or hinged panels.

(3) Height of Working Space. The work space shall be clear and extend from the grade, floor, or platform to a height of 2.0 m (6-1/2 ft) or the height of the equipment, whichever is greater. Within the height requirements of this section, other equipment that is associated with the electrical installation and is located above or below the electrical equipment shall be permitted to extend not more than 150 mm (6 in.) beyond the front of the electrical equipment.

Exception No. 1: In existing dwelling units, service equipment or panelboards that do not exceed 200 amperes shall be permitted in spaces where the height of the working space is less than 2.0 m (6-1/2 ft).

Exception No. 2: Meters that are installed in meter sockets shall be permitted to extend beyond the other equipment. The meter socket shall be required to follow the rules of this section.

(B) Clear Spaces. Working space required by this section shall not be used for storage. When normally enclosed live parts are exposed for inspection or servicing, the working space, if in a passageway or general open space, shall be suitably guarded.

(D) Illumination. Illumination shall be provided for all working spaces about service equipment, switchboards, panelboards, or motor control centers installed indoors and shall not be controlled by automatic means only. Additional lighting outlets shall not be required where the work space is illuminated by an adjacent light source or as permitted by 210.70(A)(1), Exception No. 1, for switched receptacles.

(E) Dedicated Equipment Space. All switchboards, panelboards, and motor control centers shall be located in dedicated spaces and protected from damage.

Exception: Control equipment that by its very nature or because of other rules of the Code must be adjacent to or within sight of its operating machinery shall be permitted in those locations.

(1) Indoor. Indoor installations shall comply with 110.26(E)(1)(a) through (E)(1)(d).

(a) *Dedicated Electrical Space.* The space equal to the width and depth of the equipment and extending from the floor to a height of 1.8 m (6 ft) above the equipment or to the structural ceiling, whichever is lower, shall be dedicated to the electrical installation. No piping, ducts, leak protection apparatus, or other equipment foreign to the electrical installation shall be located in this zone.

Exception: Suspended ceilings with removable panels shall be permitted within the 1.8-m (6-ft) zone.

(b) *Foreign Systems.* The area above the dedicated space required by 110.26(E)(1)(a) shall be permitted to contain foreign systems, provided protection is installed to avoid damage to the electrical equipment from condensation, leaks, or breaks in such foreign systems.

(c) *Sprinkler Protection.* Sprinkler protection shall be permitted for the dedicated space where the piping complies with this section.

(d) *Suspended Ceilings.* A dropped, suspended, or similar ceiling that does not add strength to the building structure shall not be considered a structural ceiling.

(2) Outdoor. Outdoor electrical equipment shall be installed in suitable enclosures and shall be protected from accidental contact by unauthorized personnel, or by vehicular traffic, or by accidental spillage or leakage from piping systems. The working clearance space shall include the zone described in 110.26(A). No architectural appurtenance or other equipment shall be located in this zone.

110.27 Guarding of Live Parts.

(B) Prevent Physical Damage. In locations where electrical equipment is likely to be exposed to physical damage, enclosures or guards shall be so arranged and of such strength as to prevent such damage.

110.28 Enclosure Types. Enclosures (other than surrounding fences or walls) of switchboards, panelboards, industrial control panels, motor control centers, meter sockets, enclosed switches, transfer switches, power outlets, circuit breakers, adjustable-speed drive systems, pullout switches, portable power distribution equipment, termination boxes, general-purpose transformers, fire pump controllers, fire pump motors, and motor controllers, rated not over 600 volts nominal and intended for such locations, shall be marked with an enclosure-type number as shown in Table 110.28.

Table 110.28 shall be used for selecting these enclosures for use in specific locations other than hazardous (classified) locations. The enclosures are not intended to protect against conditions such as condensation, icing, corrosion, or contamination that may occur within the enclosure or enter via the conduit or unsealed openings.

30

Table 110.28 Enclosure Selection

For Outdoor Use

Provides a Degree of Protection Against the Following Environmental Conditions	*Enclosure-Type Number* 3	3R	3S	3X	3RX	3SX	4	4X	6	6P
Incidental contact with the enclosed equipment	X	X	X	X	X	X	X	X	X	X
Rain, snow, and sleet	X	X	X	X	X	X	X	X	X	X
Sleet*	—	—	X	—	—	X	—	—	—	—
Windblown dust	X	—	X	X	X	X	X	X	X	X
Hosedown	—	—	—	—	—	—	X	X	X	X
Corrosive agents	—	—	—	X	X	X	—	X	—	X
Temporary submersion	—	—	—	—	—	—	—	—	X	X
Prolonged submersion	—	—	—	—	—	—	—	—	—	X

For Indoor Use

Provides a Degree of Protection Against the Following Environmental Conditions	*Enclosure-Type Number* 1	2	4	4X	5	6	6P	12	12K	13
Incidental contact with the enclosed equipment	X	X	X	X	X	X	X	X	X	X
Falling dirt	X	X	X	X	X	X	X	X	X	X

Falling liquids and light splashing	—	—	—	—	—	X	X	—	X X X
Circulating dust, lint, fibers, and flyings	—	X	X	X	X	X	X	—	X X X
Settling airborne dust, lint, fibers, and flyings	—	X	X	X	X	X	X	—	X X X
Hosedown and splashing water	—	—	X	X	X	X	X	—	— X X
Oil and coolant seepage	—	—	—	—	—	—	—	—	— — X
Oil or coolant spraying and splashing	—	—	—	—	—	—	—	—	— X X
Corrosive agents	—	—	—	—	—	X	X	X	X X X
Temporary submersion	—	—	—	—	X	X	X	—	— — —
Prolonged submersion	—	—	—	—	—	X	X	—	— — —

*Mechanism shall be operable when ice covered.

Informational Note No. 1: The term *raintight* is typically used in conjunction with Enclosure Types 3, 3S, 3SX, 3X, 4, 4X, 6, and 6P. The term *rainproof* is typically used in conjunction with Enclosure Types 3R, and 3RX. The term *watertight* is typically used in conjunction with Enclosure Types 4, 4X, 6, 6P. The term *driptight* is typically used in conjunction with Enclosure Types 2, 5, 12, 12K, and 13. The term *dusttight* is typically used in conjunction with Enclosure Types 3, 3S, 3SX, 3X, 5, 12, 12K, and 13.

Informational Note No. 2: Ingress protection (IP) ratings may be found in ANSI/NEMA 60529, *Degrees of Protection Provided by Enclosures*. IP ratings are not a substitute for Enclosure Type ratings.

CHAPTER 4

TEMPORARY ELECTRIC POWER

INTRODUCTION

This chapter contains provisions from Article 590—Temporary Installations. Article 590 covers provisions pertaining to temporary electric power and lighting. New construction generally requires the installation of temporary electrical equipment, which usually involves setting a pole and building a service. Although the feeders, branch circuits, and receptacles are specifically covered by Article 590, the service must be installed in accordance with Article 230 (see Chapter 6 of this book). For example, the minimum clearance of overhead service-drop conductors must comply with 230.24. Of course, not all electrical projects will require the installation of a temporary service. Remodeling and additions to existing installations typically already have available power through permanent wiring.

Because of the inherent hazards associated with construction sites, ground-fault protection of receptacles with certain ratings that are in use by personnel is required for all temporary wiring installations, not just for new construction sites with temporary services. If a receptacle(s) is installed or exists as part of the permanent wiring or the building or structure and is used for temporary electric power, ground-fault circuit-interrupter (GFCI) protection for personnel must be provided. This may require replacing the existing receptacle with a GFCI receptacle or the use of a portable GFCI device. All other outlets (besides 125-volt, single-phase, 15-, 20-, and 30-ampere receptacle outlets) must have GFCI protection for personnel, or an assured equipment grounding conductor program in accordance with 590.6(B)(2).

ARTICLE 590

Temporary Installations

590.1 Scope. The provisions of this article apply to temporary electric power and lighting installations.

590.2 All Wiring Installations.

(A) Other Articles. Except as specifically modified in this article, all other requirements of this *Code* for permanent wiring shall apply to temporary wiring installations.

(B) Approval. Temporary wiring methods shall be acceptable only if approved based on the conditions of use and any special requirements of the temporary installation.

590.3 Time Constraints.

(A) During the Period of Construction. Temporary electric power and lighting installations shall be permitted during the period of construction, remodeling, maintenance, repair, or demolition of buildings, structures, equipment, or similar activities.

(D) Removal. Temporary wiring shall be removed immediately upon completion of construction or purpose for which the wiring was installed.

590.4 General.

(A) Services. Services shall be installed in conformance with Parts I through VIII of Article 230, as applicable.

(B) Feeders. Overcurrent protection shall be provided in accordance with 240.4, 240.5, 240.100, and 240.101. Conductors shall be permitted within cable assemblies or within multiconductor cords or cables of a type identified in Table 400.4 for hard usage or extra-hard usage. For the purpose of this section, Type NM and Type NMC cables shall be permitted to be used in any dwelling, building, or structure without any height limitation or limitation by building construction type and without concealment within walls, floors, or ceilings.

(C) Branch Circuits. All branch circuits shall originate in an approved power outlet, switchboard or panelboard, motor control center, or fused switch enclosure. Conductors shall be permitted within cable assemblies or within multiconductor cord or cable of a type identified in Table 400.4 for hard usage or extra-hard usage. Conductors shall be protected from overcurrent as provided in 240.4, 240.5, and 240.100. For the purposes of this section, Type NM and Type NMC cables shall be permitted to be used in any dwelling, building, or structure without any height limitation or limitation by building construction type and without concealment within walls, floors, or ceilings.

(D) Receptacles.

(1) All Receptacles. All receptacles shall be of the grounding type. Unless installed in a continuous metal raceway that qualifies as an equipment grounding conductor in accordance with 250.118 or a continuous metal-covered cable that qualifies as an equipment grounding conductor in accordance with 250.118, all branch circuits shall include a separate equipment grounding conductor, and all receptacles shall be electrically connected to the equipment grounding conductor(s). Receptacles on construction sites shall not be installed on any branch circuit that supplies temporary lighting.

(2) Receptacles in Wet Locations. All 15- and 20-ampere, 125- and 250-volt receptacles installed in a wet location shall comply with 406.9(B)(1).

(E) Disconnecting Means. Suitable disconnecting switches or plug connectors shall be installed to permit the disconnection of all ungrounded conductors of each temporary circuit. Multiwire branch circuits shall be provided with a means to disconnect simultaneously all ungrounded conductors at the power outlet or panelboard where the branch circuit originated. Identified handle ties shall be permitted.

(F) Lamp Protection. All lamps for general illumination shall be protected from accidental contact or breakage by a suitable luminaire or lampholder with a guard.

Brass shell, paper-lined sockets, or other metal-cased sockets shall not be used unless the shell is grounded.

(G) Splices. On construction sites, a box shall not be required for splices or junction connections where the circuit conductors are multiconductor cord or cable assemblies, provided that the equipment grounding continuity is maintained with or without the box. See 110.14(B) and 400.9. A box, conduit body, or terminal fitting having a separately bushed hole for each conductor shall be used wherever a change is made to a conduit or tubing system or a metal-sheathed cable system.

(H) Protection from Accidental Damage. Flexible cords and cables shall be protected from accidental damage. Sharp corners and projections shall be avoided. Where passing through doorways or other pinch points, protection shall be provided to avoid damage.

(I) Termination(s) at Devices. Flexible cords and cables entering enclosures containing devices requiring termination shall be secured to the box with fittings designed for the purpose.

(J) Support. Cable assemblies and flexible cords and cables shall be supported in place at intervals that ensure that they will be protected from physical damage. Support shall be in the form of staples, cable ties, straps, or similar type fittings installed so as not to cause damage. Vegetation shall not be used for support of overhead spans of branch circuits or feeders.

590.5 Listing of Decorative Lighting. Decorative lighting used for holiday lighting and similar purposes, in accordance with 590.3(B), shall be listed.

590.6 Ground-Fault Protection for Personnel. Ground-fault protection for personnel for all temporary wiring installations shall be provided to comply with 590.6(A) and (B). This section shall apply only to temporary wiring installations used to supply temporary power to equipment used by personnel during construction, remodeling, maintenance, repair, or demolition of buildings, structures, equipment, or similar activities. This section shall apply to power derived from an electric utility company or from an on-site-generated power source.

(A) Receptacle Outlets. Temporary receptacle installations used to supply temporary power to equipment used by personnel during construction, remodeling, maintenance, repair, or demolition of buildings, structures, equipment, or similar activities shall comply with the requirements of 590.6(A)(1) through (A)(3), as applicable.

(1) Receptacle Outlets Not Part of Permanent Wiring. All 125-volt, single-phase, 15-, 20-, and 30-ampere receptacle outlets that are not a part of the permanent wiring of the building or structure and that are in use by personnel shall have ground-fault circuit-interrupter protection for personnel.

(2) Receptacle Outlets Existing or Installed as Permanent Wiring. Ground-fault circuit-interrupter protection for personnel shall be provided for all 125-volt, single-phase, 15-, 20-, and 30-ampere receptacle outlets installed or existing as part of the permanent wiring of the building or structure and used for temporary electric power. Listed cord sets or devices incorporating listed ground-fault circuit-interrupter protection for personnel identified for portable use shall be permitted.

(3) Receptacles on 15-kW or less Portable Generators. All 125-volt and 125/250-volt, single-phase, 15-, 20-, and 30-ampere receptacle outlets that are a part of a 15-kW or smaller portable generator shall have listed ground-fault circuit-interrupter protection for personnel. All 15- and 20-ampere, 125- and 250-volt receptacles, including those that are part of a portable generator, used in a damp or wet location shall comply with 406.9(A) and (B). Listed cord sets or devices incorporating listed ground-fault circuit-interrupter protection for personnel identified for portable use shall be permitted for use with 15-kW or less portable generators manufactured or remanufactured prior to January 1, 2011.

(B) Use of Other Outlets. For temporary wiring installations, receptacles, other than those covered by 590.6(A)(1) through (A)(3) used to supply temporary power to equipment used by personnel during construction, remodeling, maintenance, repair, or demolition of buildings, structures, or equipment, or similar activities, shall have protection in accordance with (B)(1) or the assured equipment grounding conductor program in accordance with (B)(2).

(1) GFCI Protection. Ground-fault circuit-interrupter protection for personnel.

(2) Assured Equipment Grounding Conductor Program. A written assured equipment grounding conductor program continuously enforced at the site by one or more designated persons to ensure that equipment grounding conductors for all cord sets, receptacles that are not a part of the permanent wiring of the building or structure, and equipment connected by cord and plug are installed and maintained in accordance with the applicable requirements of 250.114, 250.138, 406.4(C), and 590.4(D).

CHAPTER 5

SERVICE AND FEEDER LOAD CALCULATIONS

INTRODUCTION

Selecting the size service and panelboard(s) is one of the first steps in residential installations. Load calculations are required to size the service and/or feeder overcurrent protection, as well as the ungrounded, grounded, and grounding conductors. Article 220 contains detailed requirements for calculating branch-circuit, feeder, and service loads for one-family, two-family, and multifamily dwellings. Two load calculation methods are provided, Standard (Part III) and Optional (Part IV).

While certain installations will have a service with a single panelboard, other installations may have multiple panelboards. For dwellings with panelboards that are not part of the service, additional load calculations are necessary—one for the service and one for each feeder.

Collecting specific information and applying the appropriate demand factors will provide answers to such questions as: What is the minimum size service or feeder? What is the minimum size ungrounded (hot) conductor? What is the minimum size neutral conductor? What is the minimum size grounding electrode conductor (for services) or equipment grounding conductor (for feeders)?

ARTICLE 220

Branch-Circuit, Feeder, and Service Calculations

I. GENERAL

220.1 Scope. This article provides requirements for calculating branch-circuit, feeder, and service loads. Part I provides for general requirements for calculation methods. Part II provides calculation methods for branch-circuit loads. Parts III and IV provide calculation methods for feeders and services. Part V provides calculation methods for farms.

220.5 Calculations.

(A) Voltages. Unless other voltages are specified, for purposes of calculating branch-circuit and feeder loads, nominal system voltages of 120, 120/240, 208Y/120, 240, 347, 480Y/277, 480, 600Y/347, and 600 volts shall be used.

(B) Fractions of an Ampere. Calculations shall be permitted to be rounded to the nearest whole ampere, with decimal fractions smaller than 0.5 dropped.

II. BRANCH-CIRCUIT LOAD CALCULATIONS

220.10 General. Branch-circuit loads shall be calculated as shown in 220.12, 220.14, and 220.16.

220.12 Lighting Load for Specified Occupancies. A unit load of not less than that specified in Table 220.12 for occupancies specified therein shall constitute the minimum lighting load. The floor area for each floor shall be calculated from the outside dimensions of the building, dwelling unit, or other area involved. For dwelling units, the calculated floor area shall not include open porches, garages, or unused or unfinished spaces not adaptable for future use.

Note: Residential installations require a unit load of 33 volt-amperes per square meter (or 3 volt-amperes per square foot).

220.14 Other Loads—All Occupancies. In all occupancies, the minimum load for each outlet for general-use receptacles and outlets not used for general illumination shall not be less than that calculated in 220.14(A) through (L), the loads shown being based on nominal branch-circuit voltages.

(A) Specific Appliances or Loads. An outlet for a specific appliance or other load not covered in 220.14(B) through (L) shall be calculated based on the ampere rating of the appliance or load served.

(B) Electric Dryers and Electric Cooking Appliances in Dwelling Units. Load calculations shall be permitted as specified in 220.54 for electric dryers and in 220.55 for electric ranges and other cooking appliances.

(C) Motor Loads. Outlets for motor loads shall be calculated in accordance with the requirements in 430.22, 430.24, and 440.6.

(D) Luminaires. An outlet supplying luminaire(s) shall be calculated based on the maximum volt-ampere rating of the equipment and lamps for which the luminaire(s) is rated.

(J) Dwelling Occupancies. In one-family, two-family, and multifamily dwellings and in guest rooms or guest suites of hotels and motels, the outlets specified in (J)(1), (J)(2), and (J)(3) are included in the general lighting load calculations of 220.12. No additional load calculations shall be required for such outlets.

(1) All general-use receptacle outlets of 20-ampere rating or less, including receptacles connected to the circuits in 210.11(C)(3)
(2) The receptacle outlets specified in 210.52(E) and (G)
(3) The lighting outlets specified in 210.70(A) and (B)

220.16 Loads for Additions to Existing Installations.

(A) Dwelling Units. Loads added to an existing dwelling unit(s) shall comply with the following as applicable:

(1) Loads for structural additions to an existing dwelling unit or for a previously unwired portion of an existing dwelling unit, either of which exceeds 46.5 m² (500 ft²), shall be calculated in accordance with 220.12 and 220.14.

(2) Loads for new circuits or extended circuits in previously wired dwelling units shall be calculated in accordance with either 220.12 or 220.14, as applicable.

III. FEEDER AND SERVICE LOAD CALCULATIONS

220.40 General. The calculated load of a feeder or service shall not be less than the sum of the loads on the branch circuits supplied, as determined by Part II of this article, after any applicable demand factors permitted by Part III or IV or required by Part V have been applied.

220.42 General Lighting. The demand factors specified in Table 220.42 shall apply to that portion of the total branch-circuit load calculated for general illumination. They shall not be applied in determining the number of branch circuits for general illumination.

220.50 Motors. Motor loads shall be calculated in accordance with 430.24, 430.25, and 430.26 and with 440.6 for hermetic refrigerant motor compressors.

220.51 Fixed Electric Space Heating. Fixed electric space-heating loads shall be calculated at 100 percent of the total connected load. However, in no case shall a feeder or service load current rating be less than the rating of the largest branch circuit supplied.

220.52 Small-Appliance and Laundry Loads—Dwelling Unit.

(A) Small-Appliance Circuit Load. In each dwelling unit, the load shall be calculated at 1500 volt-amperes for each 2-wire small-appliance branch circuit as covered by 210.11(C)(1). Where the load is subdivided through two or more feeders, the calculated load for each shall include not less than 1500 volt-amperes for each 2-wire small-appliance branch circuit. These loads shall be permitted to be included with the general lighting load and subjected to the demand factors provided in Table 220.42.

Table 220.42 Lighting Load Demand Factors

Type of Occupancy	Portion of Lighting Load to Which Demand Factor Applies (Volt-Amperes)	Demand Factor (%)
Dwelling units	First 3000 or less at	100
	From 3001 to 120,000 at	35
	Remainder over 120,000 at	25

Exception: The individual branch circuit permitted by 210.52(B) (1), Exception No. 2, shall be permitted to be excluded from the calculation required by 220.52.

(B) Laundry Circuit Load. A load of not less than 1500 volt-amperes shall be included for each 2-wire laundry branch circuit installed as covered by 210.11(C)(2). This load shall be permitted to be included with the general lighting load and subjected to the demand factors provided in Table 220.42.

220.53 Appliance Load—Dwelling Unit(s). It shall be permissible to apply a demand factor of 75 percent to the nameplate rating load of four or more appliances fastened in place, other than electric ranges, clothes dryers, space-heating equipment, or air-conditioning equipment, that are served by the same feeder or service in a one-family, two-family, or multifamily dwelling.

220.54 Electric Clothes Dryers—Dwelling Unit(s). The load for household electric clothes dryers in a dwelling unit(s) shall be either 5000 watts (volt-amperes) or the nameplate rating, whichever is larger, for each dryer served. The use of the demand factors in Table 220.54 shall be permitted. Where two or more single-phase dryers are supplied by a 3-phase, 4-wire feeder or service, the total load shall be calculated on the basis of twice the maximum number connected between any two phases. Kilovolt-amperes (kVA) shall be considered equivalent to kilowatts (kW) for loads calculated in this section.

220.55 Electric Ranges and Other Cooking Appliances— Dwelling Unit(s). The load for household electric ranges, wall-mounted ovens, counter-mounted cooking units, and

Table 220.54 Demand Factors for Household Electric Clothes Dryers

Number of Dryers	Demand Factor (%)
1–4	100
5	85
6	75
7	65
8	60
9	55
10	50
11	47
12–23	47% minus 1% for each dryer exceeding 11
24–42	35% minus 0.5% for each dryer exceeding 23
43 and over	25%

other household cooking appliances individually rated in excess of 1-3/4 kW shall be permitted to be calculated in accordance with Table 220.55. Kilovolt-amperes (kVA) shall be considered equivalent to kilowatts (kW) for loads calculated under this section.

Where two or more single-phase ranges are supplied by a 3-phase, 4-wire feeder or service, the total load shall be calculated on the basis of twice the maximum number connected between any two phases.

220.60 Noncoincident Loads. Where it is unlikely that two or more noncoincident loads will be in use simultaneously, it shall be permissible to use only the largest load(s) that will be used at one time for calculating the total load of a feeder or service.

220.61 Feeder or Service Neutral Load.

(A) Basic Calculation. The feeder or service neutral load shall be the maximum unbalance of the load determined by this article. The maximum unbalanced load shall be the maximum net calculated load between the neutral conductor and any one ungrounded conductor.

Table 220.55 Demand Factors and Loads for Household Electric Ranges, Wall-Mounted Ovens, Counter-Mounted Cooking Units, and Other Household Cooking Appliances over 1-3/4 kW Rating (Column C to be used in all cases except as otherwise permitted in Note 3.)

| | Demand Factor (%) (See Notes) | | Column C |
Number of Appliances	Column A (Less than 3-1/2 kW Rating)	Column B (3-1/2 kW through 8-3/4 kW Rating)	Maximum Demand (kW) (See Notes) (Not over 12 kW Rating)
1	80	80	8
2	75	65	11
3	70	55	14
4	66	50	17
5	62	45	20
6	59	43	21
7	56	40	22
8	53	36	23
9	51	35	24
10	49	34	25
11	47	32	26
12	45	32	27
13	43	32	28

(continued)

Table 220.55 Demand Factors and Loads for Household Electric Ranges, Wall-Mounted Ovens, Counter-Mounted Cooking Units, and Other Household Cooking Appliances over 1-3/4 kW Rating (Column C to be used in all cases except as otherwise permitted in Note 3.) *Continued*

Number of Appliances	Demand Factor (%) (See Notes)		Column C Maximum Demand (kW) (See Notes) (Not over 12 kW Rating)
	Column A (Less than 3-1/2 kW Rating)	Column B (3-1/2 kW through 8-3/4 kW Rating)	
14	41	32	29
15	40	32	30
16	39	28	31
17	38	28	32
18	37	28	33
19	36	28	34
20	35	28	35
21	34	26	36
22	33	26	37
23	32	26	38
24	31	26	39
25	30	26	40

26–30	30	24	15 kW + 1 kW for each range
31–40	30	22	
41–50	30	20	
51–60	30	18	
61 and over	30	16	25 kW + 3/4 kW for each range

Notes:

1. Over 12 kW through 27 kW ranges all of same rating. For ranges individually rated more than 12 kW but not more than 27 kW, the maximum demand in Column C shall be increased 5 percent for each additional kilowatt of rating or major fraction thereof by which the rating of individual ranges exceeds 12 kW.

2. Over 8-3/4 kW through 27 kW ranges of unequal ratings. For ranges individually rated more than 8-3/4 kW and of different ratings, but none exceeding 27 kW, an average value of rating shall be calculated by adding together the ratings of all ranges to obtain the total connected load (using 12 kW for any range rated less than 12 kW) and dividing by the total number of ranges. Then the maximum demand in Column C shall be increased 5 percent for each kilowatt or major fraction thereof by which this average value exceeds 12 kW.

3. Over 1-3/4 kW through 8-3/4 kW. In lieu of the method provided in Column C, it shall be permissible to add the nameplate ratings of all household cooking appliances rated more than 1-3/4 kW but not more than 8-3/4 kW and multiply the sum by the demand factors specified in Column A or Column B for the given number of appliances. Where the rating of cooking appliances falls under both Column A and Column B, the demand factors for each column shall be applied to the appliances for that column, and the results added together.

4. Branch-Circuit Load. It shall be permissible to calculate the branch-circuit load for one range in accordance with Table 220.55. The branch-circuit load for one wall-mounted oven or one counter-mounted cooking unit shall be the nameplate rating of the appliance. The branch-circuit load for a counter-mounted cooking unit and not more than two wall-mounted ovens, all supplied from a single branch circuit and located in the same room, shall be calculated by adding the nameplate rating of the individual appliances and treating this total as equivalent to one range.

(B) Permitted Reductions. A service or feeder supplying the following loads shall be permitted to have an additional demand factor of 70 percent applied to the amount in 220.61(B)(1) or portion of the amount in 220.61(B)(2) determined by the basic calculation:

(1) A feeder or service supplying household electric ranges, wall-mounted ovens, counter-mounted cooking units, and electric dryers, where the maximum unbalanced load has been determined in accordance with Table 220.55 for ranges and Table 220.54 for dryers

(2) That portion of the unbalanced load in excess of 200 amperes where the feeder or service is supplied from a 3-wire dc or single-phase ac system; or a 4-wire, 3-phase, 3-wire, 2-phase system; or a 5-wire, 2-phase system

IV. OPTIONAL FEEDER AND SERVICE LOAD CALCULATIONS

220.82 Dwelling Unit.

(A) Feeder and Service Load. This section applies to a dwelling unit having the total connected load served by a single 120/240-volt or 208Y/120-volt set of 3-wire service or feeder conductors with an ampacity of 100 or greater. It shall be permissible to calculate the feeder and service loads in accordance with this section instead of the method specified in Part III of this article. The calculated load shall be the result of adding the loads from 220.82(B) and (C). Feeder and service-entrance conductors whose calculated load is determined by this optional calculation shall be permitted to have the neutral load determined by 220.61.

(B) General Loads. The general calculated load shall be not less than 100 percent of the first 10 kVA plus 40 percent of the remainder of the following loads:

(1) 33 volt-amperes/m^2 or 3 volt-amperes/ft^2 for general lighting and general-use receptacles. The floor area for each floor shall be calculated from the outside dimen-

sions of the dwelling unit. The calculated floor area shall not include open porches, garages, or unused or unfinished spaces not adaptable for future use.

(2) 1500 volt-amperes for each 2-wire, 20-ampere small-appliance branch circuit and each laundry branch circuit covered in 210.11(C)(1) and (C)(2).

(3) The nameplate rating of the following:

    a. All appliances that are fastened in place, permanently connected, or located to be on a specific circuit

    b. Ranges, wall-mounted ovens, counter-mounted cooking units

    c. Clothes dryers that are not connected to the laundry branch circuit specified in item (2)

    d. Water heaters

(4) The nameplate ampere or kVA rating of all permanently connected motors not included in item (3).

(C) Heating and Air-Conditioning Load. The largest of the following six selections (load in kVA) shall be included:

(1) 100 percent of the nameplate rating(s) of the air conditioning and cooling.

(2) 100 percent of the nameplate rating(s) of the heat pump when the heat pump is used without any supplemental electric heating.

(3) 100 percent of the nameplate rating(s) of the heat pump compressor and 65 percent of the supplemental electric heating for central electric space-heating systems. If the heat pump compressor is prevented from operating at the same time as the supplementary heat, it does not need to be added to the supplementary heat for the total central space heating load.

(4) 65 percent of the nameplate rating(s) of electric space heating if less than four separately controlled units.

(5) 40 percent of the nameplate rating(s) of electric space heating if four or more separately controlled units.

(6) 100 percent of the nameplate ratings of electric thermal storage and other heating systems where the usual load is expected to be continuous at the full nameplate value.

Systems qualifying under this selection shall not be calculated under any other selection in 220.82(C).

220.83 Existing Dwelling Unit. This section shall be permitted to be used to determine if the existing service or feeder is of sufficient capacity to serve additional loads. Where the dwelling unit is served by a 120/240-volt or 208Y/120-volt, 3-wire service, it shall be permissible to calculate the total load in accordance with 220.83(A) or (B).

(A) Where Additional Air-Conditioning Equipment or Electric Space-Heating Equipment Is Not to Be Installed. The following percentages shall be used for existing and additional new loads.

Load (kVA)	Percent of Load
First 8 kVA of load at	100
Remainder of load at	40

Load calculations shall include the following:

(1) General lighting and general-use receptacles at 33 volt-amperes/m² or 3 volt-amperes/ft² as determined by 220.12

(2) 1500 volt-amperes for each 2-wire, 20-ampere small-appliance branch circuit and each laundry branch circuit covered in 210.11(C)(1) and (C)(2)

(3) The nameplate rating of the following:

    a. All appliances that are fastened in place, permanently connected, or located to be on a specific circuit

    b. Ranges, wall-mounted ovens, counter-mounted cooking units

    c. Clothes dryers that are not connected to the laundry branch circuit specified in item (2)

    d. Water heaters

(B) Where Additional Air-Conditioning Equipment or Electric Space-Heating Equipment Is to Be Installed. The following percentages shall be used for existing and additional new loads. The larger connected load of air-conditioning or space-heating, but not both, shall be used.

Load	Percent of Load
Air-conditioning equipment	100
Central electric space heating	100
Less than four separately controlled space-heating units	100
First 8 kVA of all other loads	100
Remainder of all other loads	40

Other loads shall include the following:

(1) General lighting and general-use receptacles at 33 volt-amperes/m² or 3 volt-amperes/ft² as determined by 220.12
(2) 1500 volt-amperes for each 2-wire, 20-ampere small-appliance branch circuit and each laundry branch circuit covered in 210.11(C)(1) and (C)(2)
(3) The nameplate rating of the following:

    a. All appliances that are fastened in place, permanently connected, or located to be on a specific circuit
    b. Ranges, wall-mounted ovens, counter-mounted cooking units
    c. Clothes dryers that are not connected to the laundry branch circuit specified in (2)
    d. Water heaters

220.84 Multifamily Dwelling.

(A) Feeder or Service Load. It shall be permissible to calculate the load of a feeder or service that supplies three or more dwelling units of a multifamily dwelling in accordance with Table 220.84 instead of Part III of this article if all the following conditions are met:

(1) No dwelling unit is supplied by more than one feeder.
(2) Each dwelling unit is equipped with electric cooking equipment.

Exception: When the calculated load for multifamily dwellings without electric cooking in Part III of this article exceeds that calculated under Part IV for the identical load plus elec-

Table 220.84 Optional Calculations—Demand Factors for Three or More Multifamily Dwelling Units

Number of Dwelling Units	Demand Factor (%)
3–5	45
6–7	44
8–10	43
11	42
12–13	41
14–15	40
16–17	39
18–20	38
21	37
22–23	36
24–25	35
26–27	34
28–30	33
31	32
32–33	31
34–36	30
37–38	29
39–42	28
43–45	27
46–50	26
51–55	25
56–61	24
62 and over	23

tric cooking (based on 8 kW per unit), the lesser of the two loads shall be permitted to be used.

(3) Each dwelling unit is equipped with either electric space heating or air conditioning, or both. Feeders and service conductors whose calculated load is determined by this optional calculation shall be permitted to have the neutral load determined by 220.61.

(B) House Loads. House loads shall be calculated in accordance with Part III of this article and shall be in addition to the dwelling unit loads calculated in accordance with Table 220.84.

(C) Calculated Loads. The calculated load to which the demand factors of Table 220.84 apply shall include the following:

(1) 33 volt-amperes/m^2 or 3 volt-amperes/ft^2 for general lighting and general-use receptacles

(2) 1500 volt-amperes for each 2-wire, 20-ampere small-appliance branch circuit and each laundry branch circuit covered in 210.11(C)(1) and (C)(2)

(3) The nameplate rating of the following:

    a. All appliances that are fastened in place, permanently connected, or located to be on a specific circuit

    b. Ranges, wall-mounted ovens, counter-mounted cooking units

    c. Clothes dryers that are not connected to the laundry branch circuit specified in item (2)

    d. Water heaters

(4) The nameplate ampere or kVA rating of all permanently connected motors not included in item (3)

(5) The larger of the air-conditioning load or the fixed electric space-heating load

220.85 Two Dwelling Units. Where two dwelling units are supplied by a single feeder and the calculated load under Part III of this article exceeds that for three identical units calculated under 220.84, the lesser of the two loads shall be permitted to be used.

CHAPTER 6

SERVICES

INTRODUCTION

This chapter contains provisions from Article 230—Services. Article 230 covers service conductors and equipment for control and protection of services and their installation requirements. Contained in this pocket guide are seven of the eight parts that make up Article 230. The seven parts are: I General; II Overhead Service-Drop Conductors; III Underground Service-Lateral Conductors; IV Service-Entrance Conductors; V Service Equipment—General; VI Service Equipment—Disconnecting Means; and VII Service Equipment—Overcurrent Protection. Reference the 2011 *National Electrical Code* for services exceeding 600 volts, nominal.

While the service installation must comply with Parts I, IV, V, VI, and VII, compliance with Parts II and III depends on the type of service entrance. Part II contains requirements for overhead service-drop conductors, and Part III contains requirements for underground service-lateral conductors.

ARTICLE 230

Services

230.1 Scope. This article covers service conductors and equipment for control and protection of services and their installation requirements.

Informational Note: See Figure 230.1.

I. GENERAL

230.2 Number of Services. A building or other structure served shall be supplied by only one service unless permitted in 230.2(A) through (D). For the purpose of 230.40, Exception No. 2 only, underground sets of conductors, 1/0 AWG and larger, running to the same location and connected together at their supply end but not connected together at their load end shall be considered to be supplying one service.

(A) Special Conditions. Additional services shall be permitted to supply the following:

(1) Fire pumps
(2) Emergency systems
(3) Legally required standby systems
(4) Optional standby systems
(5) Parallel power production systems
(6) Systems designed for connection to multiple sources of supply for the purpose of enhanced reliability

(B) Special Occupancies. By special permission, additional services shall be permitted for either of the following:

(1) Multiple-occupancy buildings where there is no available space for service equipment accessible to all occupants
(2) A single building or other structure sufficiently large to make two or more services necessary

(C) Capacity Requirements. Additional services shall be permitted under any of the following:

(1) Where the capacity requirements are in excess of 2000 amperes at a supply voltage of 600 volts or less

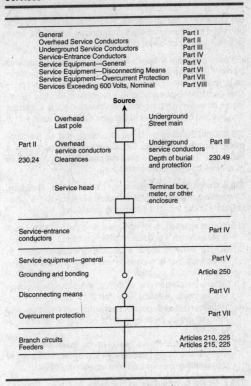

Figure 230.1 Services.

(2) Where the load requirements of a single-phase installation are greater than the serving agency normally supplies through one service

(3) By special permission

(D) Different Characteristics. Additional services shall be permitted for different voltages, frequencies, or phases, or for different uses, such as for different rate schedules.

(E) Identification. Where a building or structure is supplied by more than one service, or any combination of branch circuits, feeders, and services, a permanent plaque or directory shall be installed at each service disconnect location denoting all other services, feeders, and branch circuits supplying that building or structure and the area served by each.

230.3 One Building or Other Structure Not to Be Supplied Through Another. Service conductors supplying a building or other structure shall not pass through the interior of another building or other structure.

230.6 Conductors Considered Outside the Building. Conductors shall be considered outside of a building or other structure under any of the following conditions:

(1) Where installed under not less than 50 mm (2 in.) of concrete beneath a building or other structure
(2) Where installed within a building or other structure in a raceway that is encased in concrete or brick not less than 50 mm (2 in.) thick
(3) Where installed in any vault that meets the construction requirements of Article 450, Part III
(4) Where installed in conduit and under not less than 450 mm (18 in.) of earth beneath a building or other structure
(5) Where installed in overhead service masts on the outside surface of the building traveling through the eave of that building to meet the requirements of 230.24

230.7 Other Conductors in Raceway or Cable. Conductors other than service conductors shall not be installed in the same service raceway or service cable.

Exception No. 1: Grounding conductors and bonding jumpers.

Exception No. 2: Load management control conductors having overcurrent protection.

230.8 Raceway Seal. Where a service raceway enters a building or structure from an underground distribution system, it shall be sealed in accordance with 300.5(G). Spare or unused raceways shall also be sealed. Sealants shall be identified for use with the cable insulation, shield, or other components.

230.9 Clearances on Buildings. Service conductors and final spans shall comply with 230.9(A), (B), and (C).

(A) Clearances. Service conductors installed as open conductors or multiconductor cable without an overall outer jacket shall have a clearance of not less than 900 mm (3 ft) from windows that are designed to be opened, doors, porches, balconies, ladders, stairs, fire escapes, or similar locations.

Exception: Conductors run above the top level of a window shall be permitted to be less than the 900-mm (3-ft) requirement.

(B) Vertical Clearance. The vertical clearance of final spans above, or within 900 mm (3 ft) measured horizontally of, platforms, projections, or surfaces from which they might be reached shall be maintained in accordance with 230.24(B).

(C) Building Openings. Overhead service conductors shall not be installed beneath openings through which materials may be moved, such as openings in farm and commercial buildings, and shall not be installed where they obstruct entrance to these building openings.

230.10 Vegetation as Support. Vegetation such as trees shall not be used for support of overhead service conductors.

II. OVERHEAD SERVICE-DROP CONDUCTORS

230.22 Insulation or Covering. Individual conductors shall be insulated or covered.

Exception: The grounded conductor of a multiconductor cable shall be permitted to be bare.

230.23 Size and Rating.

(A) General. Conductors shall have sufficient ampacity to carry the current for the load as calculated in accordance with Article 220 and shall have adequate mechanical strength.

(B) Minimum Size. The conductors shall not be smaller than 8 AWG copper or 6 AWG aluminum or copper-clad aluminum.

Exception: Conductors supplying only limited loads of a single branch circuit—such as small polyphase power, controlled water heaters, and similar loads—shall not be smaller than 12 AWG hard-drawn copper or equivalent.

(C) Grounded Conductors. The grounded conductor shall not be less than the minimum size as required by 250.24(C).

230.24 Clearances. Overhead service conductors shall not be readily accessible and shall comply with 230.24(A) through (E) for services not over 600 volts, nominal.

(A) Above Roofs. Conductors shall have a vertical clearance of not less than 2.5 m (8 ft) above the roof surface. The vertical clearance above the roof level shall be maintained for a distance of not less than 900 mm (3 ft) in all directions from the edge of the roof.

Exception No. 1: The area above a roof surface subject to pedestrian or vehicular traffic shall have a vertical clearance from the roof surface in accordance with the clearance requirements of 230.24(B).

Exception No. 2: Where the voltage between conductors does not exceed 300 and the roof has a slope of 100 mm in 300 mm (4 in. in 12 in.) or greater, a reduction in clearance to 900 mm (3 ft) shall be permitted.

Exception No. 3: Where the voltage between conductors does not exceed 300, a reduction in clearance above only the overhanging portion of the roof to not less than 450 mm (18 in.) shall be permitted if (1) not more than 1.8 m (6 ft) of overhead service conductors, 1.2 m (4 ft) horizontally, pass above the roof overhang, and (2) they are terminated at a through-the-roof raceway or approved support.

Informational Note: See 230.28 for mast supports.

Exception No. 4: The requirement for maintaining the vertical clearance 900 mm (3 ft) from the edge of the roof shall not apply to the final conductor span where the service drop is attached to the side of a building.

Exception No. 5: Where the voltage between conductors does not exceed 300 and the roof area is guarded or isolated, a reduction in clearance to 900 mm (3 ft) shall be permitted.

(B) Vertical Clearance for Overhead Service Conductors. Overhead service conductors, where not in excess of 600 volts, nominal, shall have the following minimum clearance from final grade:

(1) 3.0 m (10 ft)—at the electrical service entrance to buildings, also at the lowest point of the drip loop of the building electrical entrance, and above areas or sidewalks accessible only to pedestrians, measured from final grade or other accessible surface only for service-drop cables supported on and cabled together with a grounded bare messenger where the voltage does not exceed 150 volts to ground

(2) 3.7 m (12 ft)—over residential property and driveways, and those commercial areas not subject to truck traffic where the voltage does not exceed 300 volts to ground

(3) 4.5 m (15 ft)—for those areas listed in the 3.7-m (12-ft) classification where the voltage exceeds 300 volts to ground

(4) 5.5 m (18 ft)—over public streets, alleys, roads, parking areas subject to truck traffic, driveways on other than residential property, and other land such as cultivated, grazing, forest, and orchard

(D) Clearance from Swimming Pools. See 680.8.

230.26 Point of Attachment. The point of attachment of the service-drop conductors to a building or other structure shall provide the minimum clearances as specified in 230.9 and 230.24. In no case shall this point of attachment be less than 3.0 m (10 ft) above finished grade.

230.27 Means of Attachment. Multiconductor cables used for overhead service conductors shall be attached to buildings or other structures by fittings identified for use with service conductors. Open conductors shall be attached to fittings identified for use with service conductors or to noncombustible, nonabsorbent insulators securely attached to the building or other structure.

230.28 Service Masts as Supports. Where a service mast is used for the support of service-drop conductors, it shall be of adequate strength or be supported by braces or guys to withstand safely the strain imposed by the service drop. Where raceway-type service masts are used, all raceway fittings shall be identified for use with service masts. Only power service-drop conductors shall be permitted to be attached to a service mast.

230.29 Supports over Buildings. Service conductors passing over a roof shall be securely supported by substantial structures. Where practicable, such supports shall be independent of the building.

III. UNDERGROUND SERVICE-LATERAL CONDUCTORS

230.30 Insulation. Service-lateral conductors shall be insulated for the applied voltage.

Exception: A grounded conductor shall be permitted to be uninsulated as follows:

(1) Bare copper used in a raceway.
(2) Bare copper for direct burial where bare copper is judged to be suitable for the soil conditions.
(3) Bare copper for direct burial without regard to soil conditions where part of a cable assembly identified for underground use.
(4) Aluminum or copper-clad aluminum without individual insulation or covering where part of a cable assembly identified for underground use in a raceway or for direct burial.

230.31 Size and Rating.

(A) General. Underground service conductors shall have sufficient ampacity to carry the current for the load as calculated in accordance with Article 220 and shall have adequate mechanical strength.

(B) Minimum Size. The conductors shall not be smaller than 8 AWG copper or 6 AWG aluminum or copper-clad aluminum.

Exception: Conductors supplying only limited loads of a single branch circuit—such as small polyphase power, controlled water heaters, and similar loads—shall not be smaller than 12 AWG copper or 10 AWG aluminum or copper-clad aluminum.

(C) Grounded Conductors. The grounded conductor shall not be less than the minimum size required by 250.24(C).

230.32 Protection Against Damage. Underground service conductors shall be protected against damage in accordance with 300.5. Service conductors entering a building or other structure shall be installed in accordance with 230.6 or protected by a raceway wiring method identified in 230.43.

230.33 Spliced Conductors. Service conductors shall be permitted to be spliced or tapped in accordance with 110.14, 300.5(E), 300.13, and 300.15.

IV. SERVICE-ENTRANCE CONDUCTORS

230.40 Number of Service-Entrance Conductor Sets. Each service drop, set of overhead service conductors, set of underground service conductors, or service lateral shall supply only one set of service-entrance conductors.

Exception No. 1: A building with more than one occupancy shall be permitted to have one set of service-entrance conductors for each service, as defined in 230.2, run to each occupancy or group of occupancies. If the number of service disconnect locations for any given classification of service does not exceed six, the requirements of 230.2(E) shall apply at each location. If the number of service disconnect locations exceeds six for any given supply classification, all service disconnect locations for all supply characteristics, together with any branch circuit or feeder supply sources, if applicable, shall be clearly described using suitable graphics or text, or both, on one or more plaques located in an approved, readily accessible location(s) on the building or structure served and as near as practicable to the point(s) of attachment or entry(ies) for each service drop or service lateral, and for each set of overhead or underground service conductors.

Exception No. 2: Where two to six service disconnecting means in separate enclosures are grouped at one location and supply separate loads from one service drop, set of overhead service conductors, set of underground service conductors, or service lateral, one set of service-entrance conductors shall be permitted to supply each or several such service equipment enclosures.

Exception No. 3: A single-family dwelling unit and its accessory structures shall be permitted to have one set of service-entrance conductors run to each from a single service drop, set of overhead service conductors, set of underground service conductors, or service lateral.

Exception No. 4: Two-family dwellings, multifamily dwellings, and multiple occupancy buildings shall be permitted to have one set of service-entrance conductors installed to supply the circuits covered in 210.25.

Exception No. 5: One set of service-entrance conductors connected to the supply side of the normal service disconnecting means shall be permitted to supply each or several systems covered by 230.82(5) or 230.82(6).

230.41 Insulation of Service-Entrance Conductors.
Service-entrance conductors entering or on the exterior of buildings or other structures shall be insulated.

Exception: A grounded conductor shall be permitted to be uninsulated as follows:

(1) Bare copper used in a raceway or part of a service cable assembly.
(2) Bare copper for direct burial where bare copper is judged to be suitable for the soil conditions.
(3) Bare copper for direct burial without regard to soil conditions where part of a cable assembly identified for underground use.
(4) Aluminum or copper-clad aluminum without individual insulation or covering where part of a cable assembly or identified for underground use in a raceway, or for direct burial.
(5) Bare conductors used in an auxiliary gutter.

230.42 Minimum Size and Rating.

(A) General. The ampacity of the service-entrance conductors before the application of any adjustment or correction factors shall not be less than either 230.42(A)(1) or (A)(2). Loads shall be determined in accordance with Part III, IV, or V of Article 220, as applicable. Ampacity shall be determined from 310.15. The maximum allowable current of busways shall be that value for which the busway has been listed or labeled.

(1) The sum of the noncontinuous loads plus 125 percent of continuous loads
(2) The sum of the noncontinuous load plus the continuous load if the service-entrance conductors terminate in an overcurrent device where both the overcurrent device and its assembly are listed for operation at 100 percent of their rating

(B) Specific Installations. In addition to the requirements of 230.42(A), the minimum ampacity for ungrounded conductors for specific installations shall not be less than the rating of the service disconnecting means specified in 230.79(A) through (D).

(C) Grounded Conductors. The grounded conductor shall not be smaller than the minimum size as required by 250.24(C).

230.43 Wiring Methods for 600 Volts, Nominal, or Less. Service-entrance conductors shall be installed in accordance with the applicable requirements of this *Code* covering the type of wiring method used and shall be limited to the following methods:

(1) Open wiring on insulators
(2) Type IGS cable
(3) Rigid metal conduit
(4) Intermediate metal conduit
(5) Electrical metallic tubing
(6) Electrical nonmetallic tubing (ENT)
(7) Service-entrance cables
(8) Wireways
(9) Busways
(10) Auxiliary gutters
(11) Rigid polyvinyl chloride conduit (PVC)

(12) Cablebus
(13) Type MC cable
(14) Mineral-insulated, metal-sheathed cable
(15) Flexible metal conduit not over 1.8 m (6 ft) long or liquidtight flexible metal conduit not over 1.8 m (6 ft) long between raceways, or between raceway and service equipment, with equipment bonding jumper routed with the flexible metal conduit or the liquidtight flexible metal conduit according to the provisions of 250.102(A), (B), (C), and (E)
(16) Liquidtight flexible nonmetallic conduit
(17) High density polyethylene conduit (HDPE)
(18) Nonmetallic underground conduit with conductors (NUCC)
(19) Reinforced thermosetting resin conduit (RTRC)

230.46 Spliced Conductors. Service-entrance conductors shall be permitted to be spliced or tapped in accordance with 110.14, 300.5(E), 300.13, and 300.15.

230.50 Protection Against Physical Damage.

(A) Underground Service-Entrance Conductors. Underground service-entrance conductors shall be protected against physical damage in accordance with 300.5.

(B) All Other Service-Entrance Conductors. All other service-entrance conductors, other than underground service entrance conductors, shall be protected against physical damage as specified in 230.50(B)(1) or (B)(2).

(1) Service-Entrance Cables. Service-entrance cables, where subject to physical damage, shall be protected by any of the following:

(1) Rigid metal conduit
(2) Intermediate metal conduit
(3) Schedule 80 PVC conduit
(4) Electrical metallic tubing
(5) Reinforced thermosetting resin conduit (RTRC)
(6) Other approved means

230.51 Mounting Supports. Service-entrance cables or individual open service-entrance conductors shall be supported as specified in 230.51(A), (B), or (C).

(A) Service-Entrance Cables. Service-entrance cables shall be supported by straps or other approved means within 300 mm (12 in.) of every service head, gooseneck, or connection to a raceway or enclosure and at intervals not exceeding 750 mm (30 in.).

230.53 Raceways to Drain. Where exposed to the weather, raceways enclosing service-entrance conductors shall be suitable for use in wet locations and arranged to drain. Where embedded in masonry, raceways shall be arranged to drain.

230.54 Overhead Service Locations.

(A) Service Head. Service raceways shall be equipped with a service head at the point of connection to service-drop or overhead service conductors. The service head shall be listed for use in wet locations.

(B) Service-Entrance Cables Equipped with Service Head or Gooseneck. Service-entrance cables shall be equipped with a service head. The service head shall be listed for use in wet locations.

Exception: Type SE cable shall be permitted to be formed in a gooseneck and taped with a self-sealing weather-resistant thermoplastic.

(C) Service Heads and Goosenecks Above Service-Drop or Overhead Service Attachment. Service heads and goosenecks in service-entrance cables shall be located above the point of attachment of the service-drop or overhead service conductors to the building or other structure.

Exception: Where it is impracticable to locate the service head or gooseneck above the point of attachment, the service head or gooseneck location shall be permitted not farther than 600 mm (24 in.) from the point of attachment.

(D) Secured. Service-entrance cables shall be held securely in place.

(E) Separately Bushed Openings. Service heads shall have conductors of different potential brought out through separately bushed openings.

Exception: For jacketed multiconductor service-entrance cable without splice.

(F) Drip Loops. Drip loops shall be formed on individual conductors. To prevent the entrance of moisture, service-entrance conductors shall be connected to the service-drop or overhead service conductors either (1) below the level of the service head or (2) below the level of the termination of the service-entrance cable sheath.

(G) Arranged That Water Will Not Enter Service Raceway or Equipment. Service-entrance and overhead service conductors shall be arranged so that water will not enter service raceway or equipment.

V. SERVICE EQUIPMENT—GENERAL

230.62 Service Equipment—Enclosed or Guarded. Energized parts of service equipment shall be enclosed as specified in 230.62(A) or guarded as specified in 230.62(B).

(A) Enclosed. Energized parts shall be enclosed so that they will not be exposed to accidental contact or shall be guarded as in 230.62(B).

(B) Guarded. Energized parts that are not enclosed shall be installed on a switchboard, panelboard, or control board and guarded in accordance with 110.18 and 110.27. Where energized parts are guarded as provided in 110.27(A)(1) and (A)(2), a means for locking or sealing doors providing access to energized parts shall be provided.

230.66 Marking. Service equipment rated at 600 volts or less shall be marked to identify it as being suitable for use as service equipment. All service equipment shall be listed. Individual meter socket enclosures shall not be considered service equipment.

VI. SERVICE EQUIPMENT—DISCONNECTING MEANS

230.70 General. Means shall be provided to disconnect all conductors in a building or other structure from the service-entrance conductors.

(A) Location. The service disconnecting means shall be installed in accordance with 230.70(A)(1), (A)(2), and (A)(3).

(1) Readily Accessible Location. The service disconnecting means shall be installed at a readily accessible location either outside of a building or structure or inside nearest the point of entrance of the service conductors.

(2) Bathrooms. Service disconnecting means shall not be installed in bathrooms.

(B) Marking. Each service disconnect shall be permanently marked to identify it as a service disconnect.

(C) Suitable for Use. Each service disconnecting means shall be suitable for the prevailing conditions. Service equipment installed in hazardous (classified) locations shall comply with the requirements of Articles 500 through 517.

230.71 Maximum Number of Disconnects.

(A) General. The service disconnecting means for each service permitted by 230.2, or for each set of service-entrance conductors permitted by 230.40, Exception No. 1, 3, 4, or 5, shall consist of not more than six switches or sets of circuit breakers, or a combination of not more than six switches and sets of circuit breakers, mounted in a single enclosure, in a group of separate enclosures, or in or on a switchboard. There shall be not more than six sets of disconnects per service grouped in any one location.

For the purpose of this section, disconnecting means installed as part of listed equipment and used solely for the following shall not be considered a service disconnecting means:

(1) Power monitoring equipment
(2) Surge-protective device(s)
(3) Control circuit of the ground-fault protection system
(4) Power-operable service disconnecting means

(B) Single-Pole Units. Two or three single-pole switches or breakers, capable of individual operation, shall be permitted on multiwire circuits, one pole for each ungrounded conductor, as one multipole disconnect, provided they are equipped with identified handle ties or a master handle to disconnect all conductors of the service with no more than six operations of the hand.

230.72 Grouping of Disconnects.

(A) General. The two to six disconnects as permitted in 230.71 shall be grouped. Each disconnect shall be marked to indicate the load served.

(C) Access to Occupants. In a multiple-occupancy building, each occupant shall have access to the occupant's service disconnecting means.

Exception: In a multiple-occupancy building where electric service and electrical maintenance are provided by the building management and where these are under continuous building management supervision, the service disconnecting means supplying more than one occupancy shall be permitted to be accessible to authorized management personnel only.

230.74 Simultaneous Opening of Poles. Each service disconnect shall simultaneously disconnect all ungrounded service conductors that it controls from the premises wiring system.

230.75 Disconnection of Grounded Conductor. Where the service disconnecting means does not disconnect the grounded conductor from the premises wiring, other means shall be provided for this purpose in the service equipment. A terminal or bus to which all grounded conductors can be attached by means of pressure connectors shall be permitted for this purpose. In a multisection switchboard, disconnects for the grounded conductor shall be permitted to be in any section of the switchboard, provided any such switchboard section is marked.

230.76 Manually or Power Operable. The service disconnecting means for ungrounded service conductors shall consist of one of the following:

(1) A manually operable switch or circuit breaker equipped with a handle or other suitable operating means
(2) A power-operated switch or circuit breaker, provided the switch or circuit breaker can be opened by hand in the event of a power supply failure

230.77 Indicating. The service disconnecting means shall plainly indicate whether it is in the open (off) or closed (on) position.

230.79 Rating of Service Disconnecting Means. The service disconnecting means shall have a rating not less than the calculated load to be carried, determined in accordance with Part III, IV, or V of Article 220, as applicable. In no case shall the rating be lower than specified in 230.79(A), (B), (C), or (D).

(A) One-Circuit Installations. For installations to supply only limited loads of a single branch circuit, the service disconnecting means shall have a rating of not less than 15 amperes.

(B) Two-Circuit Installations. For installations consisting of not more than two 2-wire branch circuits, the service disconnecting means shall have a rating of not less than 30 amperes.

(C) One-Family Dwellings. For a one-family dwelling, the service disconnecting means shall have a rating of not less than 100 amperes, 3-wire.

(D) All Others. For all other installations, the service disconnecting means shall have a rating of not less than 60 amperes.

230.80 Combined Rating of Disconnects. Where the service disconnecting means consists of more than one switch or circuit breaker, as permitted by 230.71, the combined ratings of all the switches or circuit breakers used shall not be less than the rating required by 230.79.

230.81 Connection to Terminals. The service conductors shall be connected to the service disconnecting means by pressure connectors, clamps, or other approved means. Connections that depend on solder shall not be used.

230.82 Equipment Connected to the Supply Side of Service Disconnect. Only the following equipment shall be permitted to be connected to the supply side of the service disconnecting means:

(1) Cable limiters or other current-limiting devices.
(2) Meters and meter sockets nominally rated not in excess of 600 volts, provided all metal housings and service

enclosures are grounded in accordance with Part VII and bonded in accordance with Part V of Article 250.

(3) Meter disconnect switches nominally rated not in excess of 600 volts that have a short-circuit current rating equal to or greater than the available short-circuit current, provided all metal housings and service enclosures are grounded in accordance with Part VII and bonded in accordance with Part V of Article 250. A meter disconnect switch shall be capable of interrupting the load served.

(4) Instrument transformers (current and voltage), impedance shunts, load management devices, surge arresters, and Type 1 surge-protective devices.

(5) Taps used only to supply load management devices, circuits for standby power systems, fire pump equipment, and fire and sprinkler alarms, if provided with service equipment and installed in accordance with requirements for service-entrance conductors.

(6) Solar photovoltaic systems, fuel cell systems, or interconnected electric power production sources.

(7) Control circuits for power-operable service disconnecting means, if suitable overcurrent protection and disconnecting means are provided.

(8) Ground-fault protection systems or Type 2 surge-protective devices, where installed as part of listed equipment, if suitable overcurrent protection and disconnecting means are provided.

VII. SERVICE EQUIPMENT— OVERCURRENT PROTECTION

230.90 Where Required. Each ungrounded service conductor shall have overload protection.

(A) Ungrounded Conductor. Such protection shall be provided by an overcurrent device in series with each ungrounded service conductor that has a rating or setting not higher than the allowable ampacity of the conductor. A set of fuses shall be considered all the fuses required to protect all the ungrounded conductors of a circuit. Single-pole circuit breakers, grouped in accordance with 230.71(B), shall be considered as one protective device.

Exception No. 2: Fuses and circuit breakers with a rating or setting that complies with 240.4(B) or (C) and 240.6 shall be permitted.

Exception No. 3: Two to six circuit breakers or sets of fuses shall be permitted as the overcurrent device to provide the overload protection. The sum of the ratings of the circuit breakers or fuses shall be permitted to exceed the ampacity of the service conductors, provided the calculated load does not exceed the ampacity of the service conductors.

Exception No. 5: Overload protection for 120/240-volt, 3-wire, single-phase dwelling services shall be permitted in accordance with the requirements of 310.15(B)(6).

(B) Not in Grounded Conductor. No overcurrent device shall be inserted in a grounded service conductor except a circuit breaker that simultaneously opens all conductors of the circuit.

230.91 Location. The service overcurrent device shall be an integral part of the service disconnecting means or shall be located immediately adjacent thereto.

230.92 Locked Service Overcurrent Devices. Where the service overcurrent devices are locked or sealed or are not readily accessible to the occupant, branch-circuit or feeder overcurrent devices shall be installed on the load side, shall be mounted in a readily accessible location, and shall be of lower ampere rating than the service overcurrent device.

230.93 Protection of Specific Circuits. Where necessary to prevent tampering, an automatic overcurrent device that protects service conductors supplying only a specific load, such as a water heater, shall be permitted to be locked or sealed where located so as to be accessible.

230.94 Relative Location of Overcurrent Device and Other Service Equipment. The overcurrent device shall protect all circuits and devices.

CHAPTER 7

FEEDERS

INTRODUCTION

This chapter contains provisions from Article 215—Feeders. Article 215 covers installation, overcurrent protection, minimum size, and ampacity of conductors for feeders that supply branch-circuit loads. A feeder consists of all circuit conductors located between the service equipment, the source of a separately derived system, or other power supply source and the final branch-circuit overcurrent device. Feeder conductors supply power to remote panelboards, which are sometimes referred to as subpanels.

Although this article does not contain specifications for calculating feeders, it does require compliance with Parts III, IV, and V of Article 220 (see Chapter 5 of this book).

ARTICLE 215

Feeders

215.1 Scope. This article covers the installation requirements, overcurrent protection requirements, minimum size, and ampacity of conductors for feeders supplying branch-circuit loads.

215.2 Minimum Rating and Size.

(A) Feeders Not More Than 600 Volts.

(1) General. Feeder conductors shall have an ampacity not less than required to supply the load as calculated in Parts III, IV, and V of Article 220. The minimum feeder-circuit conductor size, before the application of any adjustment or correction factors, shall have an allowable ampacity not less than the noncontinuous load plus 125 percent of the continuous load.

(2) Grounded Conductor. The size of the feeder circuit grounded conductor shall not be smaller than that required by 250.122, except that 250.122(F) shall not apply where grounded conductors are run in parallel.

Additional minimum sizes shall be as specified in 215.2(A)(2) and (A)(3) under the conditions stipulated.

(3) Ampacity Relative to Service Conductors. The feeder conductor ampacity shall not be less than that of the service conductors where the feeder conductors carry the total load supplied by service conductors with an ampacity of 55 amperes or less.

(4) Individual Dwelling Unit or Mobile Home Conductors. Feeder conductors for individual dwelling units or mobile homes need not be larger than service conductors. Paragraph 310.15(B)(6) shall be permitted to be used for conductor size.

Informational Note No. 2: Conductors for feeders as defined in Article 100, sized to prevent a voltage drop exceeding 3 percent at the farthest outlet of power, heating, and lighting loads, or combinations of such loads, and where the maximum total voltage drop on both feeders and branch circuits to the farthest outlet does not exceed 5 percent, will provide reasonable efficiency of operation.

215.3 Overcurrent Protection. Feeders shall be protected against overcurrent in accordance with the provisions of Part I of Article 240. Where a feeder supplies continuous loads or any combination of continuous and noncontinuous loads, the rating of the overcurrent device shall not be less than the noncontinuous load plus 125 percent of the continuous load.

215.4 Feeders with Common Neutral Conductor.

(A) Feeders with Common Neutral. Up to three sets of 3-wire feeders or two sets of 4-wire or 5-wire feeders shall be permitted to utilize a common neutral.

(B) In Metal Raceway or Enclosure. Where installed in a metal raceway or other metal enclosure, all conductors of all feeders using a common neutral conductor shall be enclosed within the same raceway or other enclosure as required in 300.20.

215.5 Diagrams of Feeders. If required by the authority having jurisdiction, a diagram showing feeder details shall be provided prior to the installation of the feeders. Such a diagram shall show the area in square feet of the building or other structure supplied by each feeder, the total calculated load before applying demand factors, the demand factors used, the calculated load after applying demand factors, and the size and type of conductors to be used.

215.6 Feeder Equipment Grounding Conductor. Where a feeder supplies branch circuits in which equipment grounding conductors are required, the feeder shall include or provide an equipment grounding conductor in accordance with the provisions of 250.134, to which the equipment grounding conductors of the branch circuits shall be connected. Where the feeder supplies a separate building or structure, the requirements of 250.32(B) shall apply.

215.9 Ground-Fault Circuit-Interrupter Protection for Personnel. Feeders supplying 15- and 20-ampere receptacle branch circuits shall be permitted to be protected by a ground-fault circuit interrupter in lieu of the provisions for such interrupters as specified in 210.8 and 590.6(A).

CHAPTER 8

PANELBOARDS AND DISCONNECTS

INTRODUCTION

This chapter contains provisions from Article 408—Switchboards and Panelboards and Article 404—Switches. Article 408 covers panelboards installed for the control of light and power circuits. A panelboard consists of one or more panel units designed for assembly in the form of a single panel. It includes buses and automatic overcurrent devices and may contain switches for the control of light, heat, or power circuits. Panelboards are accessible only from the front and are designed to be placed in cabinets or cutout boxes. Panelboards can be used as service equipment (where identified for such use) or be installed as remote panelboards (subpanels).

Provisions from Article 404 are located in two chapters of this pocket guide, Chapters 8 and 18. This chapter covers switches and circuit breakers where used as switches. The switches discussed here are more commonly called disconnect switches or safety switches. Disconnects may contain overcurrent protective devices (fuses or circuit breakers), or they may be nonfusible or molded-case switches.

ARTICLE 408

Switchboards and Panelboards

I. GENERAL

408.1 Scope. This article covers switchboards and panelboards. It does not apply to equipment operating at over 600 volts, except as specifically referenced elsewhere in the *Code.*

408.3 Support and Arrangement of Busbars and Conductors.

(C) Used as Service Equipment. Each switchboard or panelboard, if used as service equipment, shall be provided with a main bonding jumper sized in accordance with 250.28(D) or the equivalent placed within the panelboard or one of the sections of the switchboard for connecting the grounded service conductor on its supply side to the switchboard or panelboard frame. All sections of a switchboard shall be bonded together using an equipment bonding conductor sized in accordance with Table 250.122 or Table 250.66 as appropriate.

408.4 Field Identification Required.

(A) Circuit Directory or Circuit Identification. Every circuit and circuit modification shall be legibly identified as to its clear, evident, and specific purpose or use. The identification shall include sufficient detail to allow each circuit to be distinguished from all others. Spare positions that contain unused overcurrent devices or switches shall be described accordingly. The identification shall be included in a circuit directory that is located on the face or inside of the panel door in the case of a panelboard, and located at each switch or circuit breaker in a switchboard. No circuit shall be described in a manner that depends on transient conditions of occupancy.

(B) Source of Supply. All switchboards and panelboards supplied by a feeder in other than one- or two-family dwellings shall be marked to indicate the device or equipment where the power supply originates.

408.7 Unused Openings. Unused openings for circuit breakers and switches shall be closed using identified closures, or other approved means that provide protection substantially equivalent to the wall of the enclosure.

III. PANELBOARDS

408.30 General. All panelboards shall have a rating not less than the minimum feeder capacity required for the load calculated in accordance with Part III, IV, or V of Article 220, as applicable.

408.36 Overcurrent Protection. In addition to the requirement of 408.30, a panelboard shall be protected by an overcurrent protective device having a rating not greater than that of the panelboard. This overcurrent protective device shall be located within or at any point on the supply side of the panelboard.

Exception No. 1: Individual protection shall not be required for a panelboard used as service equipment with multiple disconnecting means in accordance with 230.71. In panelboards protected by three or more main circuit breakers or sets of fuses, the circuit breakers or sets of fuses shall not supply a second bus structure within the same panelboard assembly.

Exception No. 2: Individual protection shall not be required for a panelboard protected on its supply side by two main circuit breakers or two sets of fuses having a combined rating not greater than that of the panelboard. A panelboard constructed or wired under this exception shall not contain more than 42 overcurrent devices. For the purposes of determining the maximum of 42 overcurrent devices, a 2-pole or a 3-pole circuit breaker shall be considered as two or three overcurrent devices, respectively.

Exception No. 3: For existing panelboards, individual protection shall not be required for a panelboard used as service equipment for an individual residential occupancy.

(A) Snap Switches Rated at 30 Amperes or Less. Panelboards equipped with snap switches rated at 30 amperes or less shall have overcurrent protection of 200 amperes or less.

(D) Back-Fed Devices. Plug-in-type overcurrent protection devices or plug-in type main lug assemblies that are backfed and used to terminate field-installed ungrounded supply conductors shall be secured in place by an additional fastener that requires other than a pull to release the device from the mounting means on the panel.

408.37 Panelboards in Damp or Wet Locations. Panelboards in damp or wet locations shall be installed to comply with 312.2.

[**312.2 Damp and Wet Locations.** In damp or wet locations, surface-type enclosures within the scope of this article shall be placed or equipped so as to prevent moisture or water from entering and accumulating within the cabinet or cutout box, and shall be mounted so there is at least 6-mm (1/4-in.) airspace between the enclosure and the wall or other supporting surface. Enclosures installed in wet locations shall be weatherproof. For enclosures in wet locations, raceways or cables entering above the level of uninsulated live parts shall use fittings listed for wet locations.]

408.40 Grounding of Panelboards. Panelboard cabinets and panelboard frames, if of metal, shall be in physical contact with each other and shall be connected to an equipment grounding conductor. Where the panelboard is used with nonmetallic raceway or cable or where separate equipment grounding conductors are provided, a terminal bar for the equipment grounding conductors shall be secured inside the cabinet. The terminal bar shall be bonded to the cabinet and panelboard frame, if of metal; otherwise it shall be connected to the equipment grounding conductor that is run with the conductors feeding the panelboard.

Equipment grounding conductors shall not be connected to a terminal bar provided for grounded conductors or neutral conductors unless the bar is identified for the purpose and is located where interconnection between equipment grounding conductors and grounded circuit conductors is permitted or required by Article 250.

408.41 Grounded Conductor Terminations. Each grounded conductor shall terminate within the panelboard in an individual terminal that is not also used for another conductor.

Exception: Grounded conductors of circuits with parallel conductors shall be permitted to terminate in a single terminal if the terminal is identified for connection of more than one conductor.

ARTICLE 404
Switches

404.3 Enclosure.

(A) General. Switches and circuit breakers shall be of the externally operable type mounted in an enclosure listed for the intended use.

(B) Used as a Raceway. Enclosures shall not be used as junction boxes, auxiliary gutters, or raceways for conductors feeding through or tapping off to other switches or overcurrent devices, unless the enclosure complies with 312.8.

404.4 Damp or Wet Locations.

(A) Surface-Mounted Switch or Circuit Breaker. A surface-mounted switch or circuit breaker shall be enclosed in a weatherproof enclosure or cabinet that shall comply with 312.2.

[Note: See 312.2 following 408.37 in this chapter.]

(B) Flush-Mounted Switch or Circuit Breaker. A flush-mounted switch or circuit breaker shall be equipped with a weatherproof cover.

404.6 Position and Connection of Switches.

(A) Single-Throw Knife Switches. Single-throw knife switches shall be placed so that gravity will not tend to close them. Single-throw knife switches, approved for use in the inverted position, shall be provided with an integral mechanical means that ensures that the blades remain in the open position when so set.

404.7 Indicating. General-use and motor-circuit switches, circuit breakers, and molded case switches, where mounted in an enclosure as described in 404.3, shall clearly indicate whether they are in the open (off) or closed (on) position.

Where these switch or circuit breaker handles are operated vertically rather than rotationally or horizontally, the up position of the handle shall be the (on) position.

404.8 Accessibility and Grouping.

(A) Location. All switches and circuit breakers used as switches shall be located so that they may be operated from a readily accessible place. They shall be installed such that the center of the grip of the operating handle of the switch or circuit breaker, when in its highest position, is not more than 2.0 m (6 ft 7 in.) above the floor or working platform.

404.11 Circuit Breakers as Switches. A hand-operable circuit breaker equipped with a lever or handle, or a power-operated circuit breaker capable of being opened by hand in the event of a power failure, shall be permitted to serve as a switch if it has the required number of poles.

CHAPTER 9

OVERCURRENT PROTECTION

INTRODUCTION

This chapter contains provisions from Article 240—Overcurrent Protection. Article 240 provides general requirements for overcurrent protection and overcurrent protective devices. Generally, overcurrent protective devices are plug fuses, cartridge fuses, and circuit breakers. Overcurrent protective devices are usually installed within enclosures, cabinets, or cutout boxes. Automatic overcurrent devices are part of the items that make up a panelboard.

Contained in this pocket guide are six of the nine parts that are included in Article 240. The six parts are: I General; II Location; III Enclosures; IV Disconnecting and Guarding; V Plug Fuses, Fuseholders, and Adapters; and VII Circuit Breakers.

ARTICLE 240
Overcurrent Protection

I. GENERAL

240.1 Scope. Parts I through VII of this article provide the general requirements for overcurrent protection and overcurrent protective devices not more than 600 volts, nominal.

240.4 Protection of Conductors. Conductors, other than flexible cords, flexible cables, and fixture wires, shall be protected against overcurrent in accordance with their ampacities specified in 310.15, unless otherwise permitted or required in 240.4(A) through (G).

(B) Overcurrent Devices Rated 800 Amperes or Less. The next higher standard overcurrent device rating (above the ampacity of the conductors being protected) shall be permitted to be used, provided all of the following conditions are met:

(1) The conductors being protected are not part of a branch circuit supplying more than one receptacle for cord-and-plug-connected portable loads.
(2) The ampacity of the conductors does not correspond with the standard ampere rating of a fuse or a circuit breaker without overload trip adjustments above its rating (but that shall be permitted to have other trip or rating adjustments).
(3) The next higher standard rating selected does not exceed 800 amperes.

(C) Overcurrent Devices Rated over 800 Amperes. Where the overcurrent device is rated over 800 amperes, the ampacity of the conductors it protects shall be equal to or greater than the rating of the overcurrent device defined in 240.6.

(D) Small Conductors. Unless specifically permitted in 240.4(E) or (G), the overcurrent protection shall not exceed that required by (D)(1) through (D)(7) after any correction factors for ambient temperature and number of conductors have been applied.

(1) 18 AWG Copper. 7 amperes, provided all the following conditions are met:

(1) Continuous loads do not exceed 5.6 amperes.
(2) Overcurrent protection is provided by one of the following:

    a. Branch-circuit-rated circuit breakers listed and marked for use with 18 AWG copper wire

    b. Branch-circuit-rated fuses listed and marked for use with 18 AWG copper wire

    c. Class CC, Class J, or Class T fuses

(2) 16 AWG Copper. 10 amperes, provided all the following conditions are met:

(1) Continuous loads do not exceed 8 amperes.
(2) Overcurrent protection is provided by one of the following:

    a. Branch-circuit-rated circuit breakers listed and marked for use with 16 AWG copper wire

    b. Branch-circuit-rated fuses listed and marked for use with 16 AWG copper wire

    c. Class CC, Class J, or Class T fuses

(3) 14 AWG Copper. 15 amperes

(4) 12 AWG Aluminum and Copper-Clad Aluminum. 15 amperes

(5) 12 AWG Copper. 20 amperes

(6) 10 AWG Aluminum and Copper-Clad Aluminum. 25 amperes

(7) 10 AWG Copper. 30 amperes

(E) Tap Conductors. Tap conductors shall be permitted to be protected against overcurrent in accordance with the following:

(1) 210.19(A)(3) and (A)(4), Household Ranges and Cooking Appliances and Other Loads

(3) 240.21, Location in Circuit

240.6 Standard Ampere Ratings.

(A) Fuses and Fixed-Trip Circuit Breakers. The standard ampere ratings for fuses and inverse time circuit breakers shall be considered 15, 20, 25, 30, 35, 40, 45, 50, 60, 70, 80, 90, 100, 110, 125, 150, 175, 200, 225, 250, 300, 350, 400, 450, 500, 600, 700, 800, 1000, 1200, 1600, 2000, 2500, 3000, 4000, 5000, and 6000 amperes. Additional standard ampere ratings for fuses shall be 1, 3, 6, 10, and 601. The use of fuses and inverse time circuit breakers with nonstandard ampere ratings shall be permitted.

240.10 Supplementary Overcurrent Protection. Where supplementary overcurrent protection is used for luminaires, appliances, and other equipment or for internal circuits and components of equipment, it shall not be used as a substitute for required branch-circuit overcurrent devices or in place of the required branch-circuit protection. Supplementary overcurrent devices shall not be required to be readily accessible.

240.15 Ungrounded Conductors.

(A) Overcurrent Device Required. A fuse or an overcurrent trip unit of a circuit breaker shall be connected in series with each ungrounded conductor.

(B) Circuit Breaker as Overcurrent Device. Circuit breakers shall open all ungrounded conductors of the circuit both manually and automatically unless otherwise permitted in 240.15(B)(1), (B)(2), (B)(3), and (B)(4).

(1) Multiwire Branch Circuit. Individual single-pole circuit breakers, with identified handle ties, shall be permitted as the protection for each ungrounded conductor of multiwire branch circuits that serve only single-phase line-to-neutral loads.

(2) Grounded Single-Phase Alternating-Current Circuits. In grounded systems, individual single-pole circuit breakers rated 120/240 volts ac, with identified handle ties, shall be permitted as the protection for each ungrounded conductor for line-to-line connected loads for single-phase circuits.

(4) 3-Wire Direct-Current Circuits. Individual single-pole circuit breakers rated 125/250 volts dc with identified handle ties shall be permitted as the protection for each ungrounded conductor for line-to-line connected loads for 3-wire, direct-current circuits supplied from a system with a grounded neutral where the voltage to ground does not exceed 125 volts.

II. LOCATION

240.21 Location in Circuit. Overcurrent protection shall be provided in each ungrounded circuit conductor and shall be located at the point where the conductors receive their supply except as specified in 240.21(A) through (H). Conductors supplied under the provisions of 240.21(A) through (H) shall not supply another conductor except through an overcurrent protective device meeting the requirements of 240.4.

(A) Branch-Circuit Conductors. Branch-circuit tap conductors meeting the requirements specified in 210.19 shall be permitted to have overcurrent protection as specified in 210.20.

(D) Service Conductors. Service conductors shall be permitted to be protected by overcurrent devices in accordance with 230.91.

240.23 Change in Size of Grounded Conductor. Where a change occurs in the size of the ungrounded conductor, a similar change shall be permitted to be made in the size of the grounded conductor.

240.24 Location in or on Premises.

(A) Accessibility. Overcurrent devices shall be readily accessible and shall be installed so that the center of the grip of the operating handle of the switch or circuit breaker, when in its highest position, is not more than 2.0 m (6 ft 7 in.) above the floor or working platform, unless one of the following applies:

(2) For supplementary overcurrent protection, as described in 240.10.

(3) For overcurrent devices, as described in 225.40 and 230.92.

(4) For overcurrent devices adjacent to utilization equipment that they supply, access shall be permitted to be by portable means.

(B) Occupancy. Each occupant shall have ready access to all overcurrent devices protecting the conductors supplying that occupancy, unless otherwise permitted in 240.24(B) (1) and (B)(2).

(1) Service and Feeder Overcurrent Devices. Where electric service and electrical maintenance are provided by the building management and where these are under continuous building management supervision, the service overcurrent devices and feeder overcurrent devices supplying more than one occupancy shall be permitted to be accessible only to authorized management personnel in the following:

(1) Multiple-occupancy buildings
(2) Guest rooms or guest suites

(2) Branch-Circuit Overcurrent Devices. Where electric service and electrical maintenance are provided by the building management and where these are under continuous building management supervision, the branch-circuit overcurrent devices supplying any guest rooms or guest suites without permanent provisions for cooking shall be permitted to be accessible only to authorized management personnel.

(C) Not Exposed to Physical Damage. Overcurrent devices shall be located where they will not be exposed to physical damage.

(D) Not in Vicinity of Easily Ignitible Material. Overcurrent devices shall not be located in the vicinity of easily ignitible material, such as in clothes closets.

(E) Not Located in Bathrooms. In dwelling units, dormitories, and guest rooms or guest suites, overcurrent devices, other than supplementary overcurrent protection, shall not be located in bathrooms.

(F) Not Located over Steps. Overcurrent devices shall not be located over steps of a stairway.

III. ENCLOSURES

240.32 Damp or Wet Locations. Enclosures for overcurrent devices in damp or wet locations shall comply with 312.2.

[Note: See 312.2 following 408.37 in Chapter 8.]

240.33 Vertical Position. Enclosures for overcurrent devices shall be mounted in a vertical position unless that is shown to be impracticable. Circuit breaker enclosures shall be permitted to be installed horizontally where the circuit breaker is installed in accordance with 240.81.

IV. DISCONNECTING AND GUARDING

240.40 Disconnecting Means for Fuses. Cartridge fuses in circuits of any voltage where accessible to other than qualified persons, and all fuses in circuits over 150 volts to ground, shall be provided with a disconnecting means on their supply side so that each circuit containing fuses can be independently disconnected from the source of power. A current-limiting device without a disconnecting means shall be permitted on the supply side of the service disconnecting means as permitted by 230.82. A single disconnecting means shall be permitted on the supply side of more than one set of fuses as permitted by 430.112, Exception, for group operation of motors and 424.22(C) for fixed electric space-heating equipment.

V. PLUG FUSES, FUSEHOLDERS, AND ADAPTERS

240.50 General.

(A) Maximum Voltage. Plug fuses shall be permitted to be used in the following circuits:

(1) Circuits not exceeding 125 volts between conductors
(2) Circuits supplied by a system having a grounded neutral point where the line-to-neutral voltage does not exceed 150 volts

(B) Marking. Each fuse, fuseholder, and adapter shall be marked with its ampere rating.

(C) Hexagonal Configuration. Plug fuses of 15-ampere and lower rating shall be identified by a hexagonal configuration of the window, cap, or other prominent part to distinguish them from fuses of higher ampere ratings.

(D) No Energized Parts. Plug fuses, fuseholders, and adapters shall have no exposed energized parts after fuses or fuses and adapters have been installed.

(E) Screw Shell. The screw shell of a plug-type fuseholder shall be connected to the load side of the circuit.

240.51 Edison-Base Fuses.

(A) Classification. Plug fuses of the Edison-base type shall be classified at not over 125 volts and 30 amperes and below.

(B) Replacement Only. Plug fuses of the Edison-base type shall be used only for replacements in existing installations where there is no evidence of overfusing or tampering.

240.52 Edison-Base Fuseholders.
Fuseholders of the Edison-base type shall be installed only where they are made to accept Type S fuses by the use of adapters.

240.53 Type S Fuses.
Type S fuses shall be of the plug type and shall comply with 240.53(A) and (B).

(A) Classification. Type S fuses shall be classified at not over 125 volts and 0 to 15 amperes, 16 to 20 amperes, and 21 to 30 amperes.

(B) Noninterchangeable. Type S fuses of an ampere classification as specified in 240.53(A) shall not be interchangeable with a lower ampere classification. They shall be designed so that they cannot be used in any fuseholder other than a Type S fuseholder or a fuseholder with a Type S adapter inserted.

240.54 Type S Fuses, Adapters, and Fuseholders.

(A) To Fit Edison-Base Fuseholders. Type S adapters shall fit Edison-base fuseholders.

(B) To Fit Type S Fuses Only. Type S fuseholders and adapters shall be designed so that either the fuseholder itself or the fuseholder with a Type S adapter inserted cannot be used for any fuse other than a Type S fuse.

(C) Nonremovable. Type S adapters shall be designed so that once inserted in a fuseholder, they cannot be removed.

(D) Nontamperable. Type S fuses, fuseholders, and adapters shall be designed so that tampering or shunting (bridging) would be difficult.

(E) Interchangeability. Dimensions of Type S fuses, fuseholders, and adapters shall be standardized to permit interchangeability regardless of the manufacturer.

VII. CIRCUIT BREAKERS

240.80 Method of Operation. Circuit breakers shall be trip free and capable of being closed and opened by manual operation. Their normal method of operation by other than manual means, such as electrical or pneumatic, shall be permitted if means for manual operation are also provided.

240.81 Indicating. Circuit breakers shall clearly indicate whether they are in the open "off" or closed "on" position.

Where circuit breaker handles are operated vertically rather than rotationally or horizontally, the "up" position of the handle shall be the "on" position.

240.83 Marking.

(A) Durable and Visible. Circuit breakers shall be marked with their ampere rating in a manner that will be durable and visible after installation. Such marking shall be permitted to be made visible by removal of a trim or cover.

(B) Location. Circuit breakers rated at 100 amperes or less and 600 volts or less shall have the ampere rating molded, stamped, etched, or similarly marked into their handles or escutcheon areas.

(C) Interrupting Rating. Every circuit breaker having an interrupting rating other than 5000 amperes shall have its interrupting rating shown on the circuit breaker. The interrupting rating shall not be required to be marked on circuit breakers used for supplementary protection.

CHAPTER 10

GROUNDED (NEUTRAL) CONDUCTORS

INTRODUCTION

This chapter contains provisions from Article 200—Use and Identification of Grounded Conductors. Article 200 covers grounded conductors in premises wiring systems and identification of grounded conductors and their terminals. A grounded conductor is a system or circuit conductor that is intentionally grounded. Grounded conductors are generally identified by a continuous white or gray outer finish or by three continuous white stripes on other than green insulation along its entire length. Although this is a short article, do not overlook the importance of proper use and identification of grounded conductors.

ARTICLE 200
Use and Identification of Grounded Conductors

200.1 Scope. This article provides requirements for the following:

(1) Identification of terminals
(2) Grounded conductors in premises wiring systems
(3) Identification of grounded conductors

200.2 General. Grounded conductors shall comply with 200.2(A) and (B).

(A) Insulation. The grounded conductor, where insulated, shall have insulation that is (1) suitable, other than color, for any ungrounded conductor of the same circuit on circuits of less than 1000 volts or impedance grounded neutral systems of 1 kV and over, or (2) rated not less than 600 volts for solidly grounded neutral systems of 1 kV and over as described in 250.184(A).

(B) Continuity. The continuity of a grounded conductor shall not depend on a connection to a metallic enclosure, raceway, or cable armor.

200.3 Connection to Grounded System. Premises wiring shall not be electrically connected to a supply system unless the latter contains, for any grounded conductor of the interior system, a corresponding conductor that is grounded. For the purpose of this section, *electrically connected* shall mean connected so as to be capable of carrying current, as distinguished from connection through electromagnetic induction.

200.4 Neutral Conductors. Neutral conductors shall not be used for more than one branch circuit, for more than one multiwire branch circuit, or for more than one set of ungrounded feeder conductors unless specifically permitted elsewhere in this *Code*.

200.6 Means of Identifying Grounded Conductors.

(A) Sizes 6 AWG or Smaller. An insulated grounded conductor of 6 AWG or smaller shall be identified by one of the following means:

(1) A continuous white outer finish.

(2) A continuous gray outer finish.

(3) Three continuous white stripes along the conductor's entire length on other than green insulation.

(4) Wires that have their outer covering finished to show a white or gray color but have colored tracer threads in the braid identifying the source of manufacture shall be considered as meeting the provisions of this section.

(B) Sizes 4 AWG or Larger. An insulated grounded conductor 4 AWG or larger shall be identified by one of the following means:

(1) A continuous white outer finish.

(2) A continuous gray outer finish

(3) Three continuous white stripes along its entire length on other than green insulation.

(4) At the time of installation, by a distinctive white or gray marking at its terminations. This marking shall encircle the conductor or insulation.

(C) Flexible Cords. An insulated conductor that is intended for use as a grounded conductor, where contained within a flexible cord, shall be identified by a white or gray outer finish or by methods permitted by 400.22.

(E) Grounded Conductors of Multiconductor Cables. The insulated grounded conductors in a multiconductor cable shall be identified by a continuous white or gray outer finish or by three continuous white stripes on other than green insulation along its entire length. Multiconductor flat cable 4 AWG or larger shall be permitted to employ an external ridge on the grounded conductor.

200.7 Use of Insulation of a White or Gray Color or with Three Continuous White Stripes.

(A) General. The following shall be used only for the grounded circuit conductor, unless otherwise permitted in 200.7(B) and (C):

(1) A conductor with continuous white or gray covering

(2) A conductor with three continuous white stripes on other than green insulation

(3) A marking of white or gray color at the termination

(B) Circuits of Less Than 50 Volts. A conductor with white or gray color insulation or three continuous white stripes or having a marking of white or gray at the termination for circuits of less than 50 volts shall be required to be grounded only as required by 250.20(A).

(C) Circuits of 50 Volts or More. The use of insulation that is white or gray or that has three continuous white stripes for other than a grounded conductor for circuits of 50 volts or more shall be permitted only as in (1) and (2).

(1) If part of a cable assembly that has the insulation permanently reidentified to indicate its use as an ungrounded conductor by marking tape, painting, or other effective means at its termination and at each location where the conductor is visible and accessible. Identification shall encircle the insulation and shall be a color other than white, gray, or green. If used for single-pole, 3-way or 4-way switch loops, the reidentified conductor with white or gray insulation or three continuous white stripes shall be used only for the supply to the switch, but not as a return conductor from the switch to the outlet.

(2) A flexible cord, having one conductor identified by a white or gray outer finish or three continuous white stripes or by any other means permitted by 400.22, that is used for connecting an appliance or equipment permitted by 400.7. This shall apply to flexible cords connected to outlets whether or not the outlet is supplied by a circuit that has a grounded conductor.

200.9 Means of Identification of Terminals. The identification of terminals to which a grounded conductor is to be connected shall be substantially white in color. The identification of other terminals shall be of a readily distinguishable different color.

200.11 Polarity of Connections. No grounded conductor shall be attached to any terminal or lead so as to reverse the designated polarity.

CHAPTER 11

GROUNDING AND BONDING

INTRODUCTION

This chapter contains provisions from Article 250—Grounding and Bonding. Article 250 covers general requirements for grounding and bonding of electrical installations: systems, circuits, equipment, and conductors. Having a good understanding of words and terms within Article 250 is imperative. Review the related terminology in Chapter 2 of this book, including bonding (bonded), bonding jumper, equipment bonding jumper, main bonding jumper, ground, grounded, grounded conductor, grounding conductor, equipment grounding conductor, and grounding electrode conductor.

Contained in this pocket guide are seven of the ten parts that are included in Article 250. Those seven parts are: I General; II Circuit System Grounding; III Grounding Electrode System and Grounding Electrode Conductor; IV Enclosure, Raceway, and Service Cable Grounding; V Bonding; VI Equipment Grounding and Equipment Grounding Conductors; and VII Methods of Equipment Grounding.

ARTICLE 250
Grounding and Bonding

I. GENERAL

250.1 Scope. This article covers general requirements for grounding and bonding of electrical installations, and the specific requirements in (1) through (6).

(1) Systems, circuits, and equipment required, permitted, or not permitted to be grounded
(2) Circuit conductor to be grounded on grounded systems
(3) Location of grounding connections
(4) Types and sizes of grounding and bonding conductors and electrodes
(5) Methods of grounding and bonding
(6) Conditions under which guards, isolation, or insulation may be substituted for grounding

250.4 General Requirements for Grounding and Bonding. The following general requirements identify what grounding and bonding of electrical systems are required to accomplish. The prescriptive methods contained in Article 250 shall be followed to comply with the performance requirements of this section.

(A) Grounded Systems.

(1) Electrical System Grounding. Electrical systems that are grounded shall be connected to earth in a manner that will limit the voltage imposed by lightning, line surges, or unintentional contact with higher-voltage lines and that will stabilize the voltage to earth during normal operation.

(2) Grounding of Electrical Equipment. Normally non–current-carrying conductive materials enclosing electrical conductors or equipment, or forming part of such equipment, shall be connected to earth so as to limit the voltage to ground on these materials.

(3) Bonding of Electrical Equipment. Normally non–current-carrying conductive materials enclosing electrical conductors or equipment, or forming part of such equipment, shall be connected together and to the electrical supply source in a manner that establishes an effective ground-fault current path.

(4) Bonding of Electrically Conductive Materials and Other Equipment. Normally non–current-carrying electrically conductive materials that are likely to become energized shall be connected together and to the electrical supply source in a manner that establishes an effective ground-fault current path.

(5) Effective Ground-Fault Current Path. Electrical equipment and wiring and other electrically conductive material likely to become energized shall be installed in a manner that creates a low-impedance circuit facilitating the operation of the overcurrent device or ground detector for high-impedance grounded systems. It shall be capable of safely carrying the maximum ground-fault current likely to be imposed on it from any point on the wiring system where a ground fault may occur to the electrical supply source. The earth shall not be considered as an effective ground-fault current path.

250.6 Objectionable Current.

(A) Arrangement to Prevent Objectionable Current. The grounding of electrical systems, circuit conductors, surge arresters, surge-protective devices, and conductive normally non–current-carrying metal parts of equipment shall be installed and arranged in a manner that will prevent objectionable current.

(B) Alterations to Stop Objectionable Current. If the use of multiple grounding connections results in objectionable current, one or more of the following alterations shall be permitted to be made, provided that the requirements of 250.4(A)(5) or (B)(4) are met:

(1) Discontinue one or more but not all of such grounding connections.
(2) Change the locations of the grounding connections.
(3) Interrupt the continuity of the conductor or conductive path causing the objectionable current.
(4) Take other suitable remedial and approved action.

250.8 Connection of Grounding and Bonding Equipment.

(A) Permitted Methods. Equipment grounding conductors, grounding electrode conductors, and bonding jumpers shall be connected by one of the following means:

(1) Listed pressure connectors
(2) Terminal bars
(3) Pressure connectors listed as grounding and bonding equipment
(4) Exothermic welding process
(5) Machine screw-type fasteners that engage not less than two threads or are secured with a nut
(6) Thread-forming machine screws that engage not less than two threads in the enclosure
(7) Connections that are part of a listed assembly
(8) Other listed means

(B) Methods Not Permitted. Connection devices or fittings that depend solely on solder shall not be used.

250.10 Protection of Ground Clamps and Fittings. Ground clamps or other fittings shall be approved for general use without protection or shall be protected from physical damage as indicated in (1) or (2) as follows:

(1) In installations where they are not likely to be damaged
(2) Where enclosed in metal, wood, or equivalent protective covering

250.12 Clean Surfaces. Nonconductive coatings (such as paint, lacquer, and enamel) on equipment to be grounded shall be removed from threads and other contact surfaces to ensure good electrical continuity or be connected by means of fittings designed so as to make such removal unnecessary.

II. SYSTEM GROUNDING

250.20 Alternating-Current Systems to Be Grounded. Alternating-current systems shall be grounded as provided for in 250.20(A), (B), (C), or (D). Other systems shall be permitted to be grounded. If such systems are grounded, they shall comply with the applicable provisions of this article.

(A) Alternating-Current Systems of Less Than 50 Volts.
Alternating-current systems of less than 50 volts shall be grounded under any of the following conditions:

(1) Where supplied by transformers, if the transformer supply system exceeds 150 volts to ground
(2) Where supplied by transformers, if the transformer supply system is ungrounded
(3) Where installed outside as overhead conductors

(B) Alternating-Current Systems of 50 Volts to 1000 Volts. Alternating-current systems of 50 volts to less than 1000 volts that supply premises wiring and premises wiring systems shall be grounded under any of the following conditions:

(1) Where the system can be grounded so that the maximum voltage to ground on the ungrounded conductors does not exceed 150 volts
(2) Where the system is 3-phase, 4-wire, wye connected in which the neutral conductor is used as a circuit conductor
(3) Where the system is 3-phase, 4-wire, delta connected in which the midpoint of one phase winding is used as a circuit conductor

250.24 Grounding Service-Supplied Alternating-Current Systems.

(A) System Grounding Connections. A premises wiring system supplied by a grounded ac service shall have a grounding electrode conductor connected to the grounded service conductor, at each service, in accordance with 250.24(A)(1) through (A)(5).

(1) General. The grounding electrode conductor connection shall be made at any accessible point from the load end of the service drop or service lateral to and including the terminal or bus to which the grounded service conductor is connected at the service disconnecting means.

(4) Main Bonding Jumper as Wire or Busbar. Where the main bonding jumper specified in 250.28 is a wire or busbar and is installed from the grounded conductor terminal bar or bus to the equipment grounding terminal bar or bus in the service equipment, the grounding electrode conductor shall be permitted to be connected to the equipment grounding terminal, bar, or bus to which the main bonding jumper is connected.

(5) Load-Side Grounding Connections. A grounded conductor shall not be connected to normally non–current-carrying metal parts of equipment, to equipment grounding conductor(s), or be reconnected to ground on the load side of the service disconnecting means except as otherwise permitted in this article.

(B) Main Bonding Jumper. For a grounded system, an unspliced main bonding jumper shall be used to connect the equipment grounding conductor(s) and the service-disconnect enclosure to the grounded conductor within the enclosure for each service disconnect in accordance with 250.28.

Exception No. 1: Where more than one service disconnecting means is located in an assembly listed for use as service equipment, an unspliced main bonding jumper shall bond the grounded conductor(s) to the assembly enclosure.

(C) Grounded Conductor Brought to Service Equipment. Where an ac system operating at less than 1000 volts is grounded at any point, the grounded conductor(s) shall be routed with the ungrounded conductors to each service disconnecting means and shall be connected to each disconnecting means grounded conductor(s) terminal or bus. A main bonding jumper shall connect the grounded conductor(s) to each service disconnecting means enclosure. The grounded conductor(s) shall be installed in accordance with 250.24(C)(1) through (C)(4).

(1) Sizing for a Single Raceway. The grounded conductor shall not be smaller than the required grounding electrode conductor specified in Table 250.66 but shall not be required to be larger than the largest ungrounded service-entrance conductor(s). In addition, for sets of ungrounded service-entrance conductors larger than 1100 kcmil copper or 1750 kcmil aluminum, the grounded conductor shall not be smaller than 12-1/2 percent of the circular mil area of the largest set of service-entrance ungrounded conductor(s).

(2) Parallel Conductors in Two or More Raceways. If the ungrounded service-entrance conductors are installed in parallel in two or more raceways, the grounded conductor shall also be installed in parallel. The size of the grounded conductor in each raceway shall be based on the total circular mil area of the parallel ungrounded conductors in the raceway, as indicated in 250.24(C)(1), but not smaller than 1/0 AWG.

Informational Note: See 310.10(H) for grounded conductors connected in parallel.

(D) Grounding Electrode Conductor. A grounding electrode conductor shall be used to connect the equipment grounding conductors, the service-equipment enclosures, and, where the system is grounded, the grounded service conductor to the grounding electrode(s) required by Part III of this article. This conductor shall be sized in accordance with 250.66.

250.26 Conductor to Be Grounded—Alternating-Current Systems. For ac premises wiring systems, the conductor to be grounded shall be as specified in the following:

(1) Single-phase, 2-wire—one conductor
(2) Single-phase, 3-wire—the neutral conductor
(3) Multiphase systems having one wire common to all phases—the common conductor
(4) Multiphase systems where one phase is grounded—one phase conductor
(5) Multiphase systems in which one phase is used as in (2)—the neutral conductor

250.28 Main Bonding Jumper and System Bonding Jumper. For a grounded system, main bonding jumpers and system bonding jumpers shall be installed as follows:

(A) Material. Main bonding jumpers and system bonding jumpers shall be of copper or other corrosion-resistant material. A main bonding jumper and a system bonding jumper shall be a wire, bus, screw, or similar suitable conductor.

(B) Construction. Where a main bonding jumper or a system bonding jumper is a screw only, the screw shall be identified with a green finish that shall be visible with the screw installed.

(C) Attachment. Main bonding jumpers and system bonding jumpers shall be connected in the manner specified by the applicable provisions of 250.8.

(D) Size. Main bonding jumpers and system bonding jumpers shall be sized in accordance with 250.28(D)(1) through (D)(3).

(1) General. Main bonding jumpers and system bonding jumpers shall not be smaller than the sizes shown in Table 250.66. Where the supply conductors are larger than 1100 kcmil copper or 1750 kcmil aluminum, the bonding jumper shall have an area that is not less than 12-1/2 percent of the area of the largest phase conductor except that, where the phase conductors and the bonding jumper are of different materials (copper or aluminum), the minimum size of the bonding jumper shall be based on the assumed use of phase conductors of the same material as the bonding jumper and with an ampacity equivalent to that of the installed phase conductors.

(2) Main Bonding Jumper for Service with More Than One Enclosure. Where a service consists of more than a single enclosure as permitted in 230.71(A), the main bonding jumper for each enclosure shall be sized in accordance with 250.28(D)(1) based on the largest ungrounded service conductor serving that enclosure.

250.32 Buildings or Structures Supplied by a Feeder(s) or Branch Circuit(s). See 250.32 in Chapter 24

III. GROUNDING ELECTRODE SYSTEM AND GROUNDING ELECTRODE CONDUCTOR

250.50 Grounding Electrode System. All grounding electrodes as described in 250.52(A)(1) through (A)(7) that are present at each building or structure served shall be bonded together to form the grounding electrode system. Where none of these grounding electrodes exist, one or more of the grounding electrodes specified in 250.52(A)(4) through (A)(8) shall be installed and used.

Exception: Concrete-encased electrodes of existing buildings or structures shall not be required to be part of the grounding electrode system where the steel reinforcing bars or rods are not accessible for use without disturbing the concrete.

250.52 Grounding Electrodes.

(A) Electrodes Permitted for Grounding.

(1) Metal Underground Water Pipe. A metal underground water pipe in direct contact with the earth for 3.0 m (10 ft) or more (including any metal well casing bonded to the pipe) and electrically continuous (or made electrically continuous by bonding around insulating joints or insulating pipe) to the points of connection of the grounding electrode conductor and the bonding conductor(s) or jumper(s), if installed.

(2) Metal Frame of the Building or Structure. The metal frame of the building or structure that is connected to the earth by one or more of the following methods:

(1) At least one structural metal member that is in direct contact with the earth for 3.0 m (10 ft) or more, with or without concrete encasement.
(2) Hold-down bolts securing the structural steel column that are connected to a concrete-encased electrode that complies with 250.52(A)(3) and is located in the support footing or foundation. The hold-down bolts shall be connected to the concrete-encased electrode by welding, exothermic welding, the usual steel tie wires, or other approved means.

(3) Concrete-Encased Electrode. A concrete-encased electrode shall consist of at least 6.0 m (20 ft) of either (1) or (2):

(1) One or more bare or zinc galvanized or other electrically conductive coated steel reinforcing bars or rods of not less than 13 mm (1/2 in.) in diameter, installed in one

continuous 6.0 m (20 ft) length, or if in multiple pieces connected together by the usual steel tie wires, exothermic welding, welding, or other effective means to create a 6.0 m (20 ft) or greater length; or

(2) Bare copper conductor not smaller than 4 AWG

Metallic components shall be encased by at least 50 mm (2 in.) of concrete and shall be located horizontally within that portion of a concrete foundation or footing that is in direct contact with the earth or within vertical foundations or structural components or members that are in direct contact with the earth. If multiple concrete-encased electrodes are present at a building or structure, it shall be permissible to bond only one into the grounding electrode system.

Informational Note: Concrete installed with insulation, vapor barriers, films or similar items separating the concrete from the earth is not considered to be in "direct contact" with the earth.

(4) Ground Ring. A ground ring encircling the building or structure, in direct contact with the earth, consisting of at least 6.0 m (20 ft) of bare copper conductor not smaller than 2 AWG.

(5) Rod and Pipe Electrodes. Rod and pipe electrodes shall not be less than 2.44 m (8 ft) in length and shall consist of the following materials.

(a) Grounding electrodes of pipe or conduit shall not be smaller than metric designator 21 (trade size 3/4) and, where of steel, shall have the outer surface galvanized or otherwise metal-coated for corrosion protection.

(b) Rod-type grounding electrodes of stainless steel and copper or zinc coated steel shall be at least 15.87 mm (5/8 in.) in diameter, unless listed.

(6) Other Listed Electrodes. Other listed grounding electrodes shall be permitted.

(7) Plate Electrodes. Each plate electrode shall expose not less than 0.186 m^2 (2 ft^2) of surface to exterior soil. Electrodes of bare or conductively coated iron or steel plates shall be at least 6.4 mm (1/4 in.) in thickness. Solid, uncoated electrodes of nonferrous metal shall be at least 1.5 mm (0.06 in.) in thickness.

(8) Other Local Metal Underground Systems or Structures. Other local metal underground systems or structures such as piping systems, underground tanks, and underground metal well casings that are not bonded to a metal water pipe.

(B) Not Permitted for Use as Grounding Electrodes. The following systems and materials shall not be used as grounding electrodes:

(1) Metal underground gas piping systems
(2) Aluminum

250.53 Grounding Electrode System Installation.

(A) Rod, Pipe, and Plate Electrodes. Rod, pipe, and plate electrodes shall meet the requirements of 250.53(A) (1) through (A)(3).

(1) Below Permanent Moisture Level. If practicable, rod, pipe, and plate electrodes shall be embedded below permanent moisture level. Rod, pipe, and plate electrodes shall be free from nonconductive coatings such as paint or enamel.

(2) Supplemental Electrode Required. A single rod, pipe, or plate electrode shall be supplemented by an additional electrode of a type specified in 250.52(A)(2) through (A)(8). The supplemental electrode shall be permitted to be bonded to one of the following:

(1) Rod, pipe, or plate electrode
(2) Grounding electrode conductor
(3) Grounded service-entrance conductor
(4) Nonflexible grounded service raceway
(5) Any grounded service enclosure

Exception: If a single rod, pipe, or plate grounding electrode has a resistance to earth of 25 ohms or less, the supplemental electrode shall not be required.

(3) Supplemental Electrode. If multiple rod, pipe, or plate electrodes are installed to meet the requirements of this section, they shall not be less than 1.8 m (6 ft) apart.

Informational Note: The paralleling efficiency of rods is increased by spacing them twice the length of the longest rod.

(B) Electrode Spacing. Where more than one of the electrodes of the type specified in 250.52(A)(5) or (A)(7) are used, each electrode of one grounding system (including that used for strike termination devices) shall not be less than 1.83 m (6 ft) from any other electrode of another grounding system. Two or more grounding electrodes that are bonded together shall be considered a single grounding electrode system.

(C) Bonding Jumper. The bonding jumper(s) used to connect the grounding electrodes together to form the grounding electrode system shall be installed in accordance with 250.64(A), (B), and (E), shall be sized in accordance with 250.66, and shall be connected in the manner specified in 250.70.

(D) Metal Underground Water Pipe. If used as a grounding electrode, metal underground water pipe shall meet the requirements of 250.53(D)(1) and (D)(2).

(1) Continuity. Continuity of the grounding path or the bonding connection to interior piping shall not rely on water meters or filtering devices and similar equipment.

(2) Supplemental Electrode Required. A metal underground water pipe shall be supplemented by an additional electrode of a type specified in 250.52(A)(2) through (A)(8). If the supplemental electrode is of the rod, pipe, or plate type, it shall comply with 250.53(A). The supplemental electrode shall be bonded to one of the following:

(1) Grounding electrode conductor
(2) Grounded service-entrance conductor
(3) Nonflexible grounded service raceway
(4) Any grounded service enclosure
(5) As provided by 250.32(B)

Exception: The supplemental electrode shall be permitted to be bonded to the interior metal water piping at any convenient point as specified in 250.68(C)(1), Exception.

(E) Supplemental Electrode Bonding Connection Size. Where the supplemental electrode is a rod, pipe, or plate electrode, that portion of the bonding jumper that is the sole connection to the supplemental grounding electrode shall not be required to be larger than 6 AWG copper wire or 4 AWG aluminum wire.

(F) Ground Ring. The ground ring shall be buried at a depth below the earth's surface of not less than 750 mm (30 in.).

(G) Rod and Pipe Electrodes. The electrode shall be installed such that at least 2.44 m (8 ft) of length is in contact with the soil. It shall be driven to a depth of not less than 2.44 m (8 ft) except that, where rock bottom is encountered, the electrode shall be driven at an oblique angle not to exceed 45 degrees from the vertical or, where rock bottom is encountered at an angle up to 45 degrees, the electrode shall be permitted to be buried in a trench that is at least 750 mm (30 in.) deep. The upper end of the electrode shall be flush with or below ground level unless the aboveground end and the grounding electrode conductor attachment are protected against physical damage as specified in 250.10.

(H) Plate Electrode. Plate electrodes shall be installed not less than 750 mm (30 in.) below the surface of the earth.

250.54 Auxiliary Grounding Electrodes. One or more grounding electrodes shall be permitted to be connected to the equipment grounding conductors specified in 250.118 and shall not be required to comply with the electrode bonding requirements of 250.50 or 250.53(C) or the resistance requirements of 250.53(A)(2) Exception, but the earth shall not be used as an effective ground-fault current path as specified in 250.4(A)(5) and 250.4(B)(4).

250.58 Common Grounding Electrode. Where an ac system is connected to a grounding electrode in or at a building or structure, the same electrode shall be used to ground conductor enclosures and equipment in or on that building or structure. Where separate services, feeders, or branch circuits supply a building and are required to be connected to a grounding electrode(s), the same grounding electrode(s) shall be used.

Two or more grounding electrodes that are bonded together shall be considered as a single grounding electrode system in this sense.

250.60 Use of Strike Termination Devices. Conductors and driven pipes, rods, or plate electrodes used for grounding strike termination devices shall not be used in lieu of the grounding electrodes required by 250.50 for grounding wiring systems and equipment. This provision shall not prohibit the required bonding together of grounding electrodes of different systems.

250.62 Grounding Electrode Conductor Material. The grounding electrode conductor shall be of copper, aluminum, or copper-clad aluminum. The material selected shall be resistant to any corrosive condition existing at the installation or shall be protected against corrosion. The conductor shall be solid or stranded, insulated, covered, or bare.

250.64 Grounding Electrode Conductor Installation. Grounding electrode conductors at the service, at each building or structure where supplied by a feeder(s) or branch circuit(s), or at a separately derived system shall be installed as specified in 250.64(A) through (F).

(A) Aluminum or Copper-Clad Aluminum Conductors. Bare aluminum or copper-clad aluminum grounding electrode conductors shall not be used where in direct contact with masonry or the earth or where subject to corrosive conditions. Where used outside, aluminum or copper-clad aluminum grounding electrode conductors shall not be terminated within 450 mm (18 in.) of the earth.

(B) Securing and Protection Against Physical Damage. Where exposed, a grounding electrode conductor or its enclosure shall be securely fastened to the surface on which it is carried. Grounding electrode conductors shall be permitted to be installed on or through framing members. A 4 AWG or larger copper or aluminum grounding electrode conductor shall be protected if exposed to physical damage. A 6 AWG grounding electrode conductor that is free from exposure to physical damage shall be permitted to be run along the surface of the building construction without metal covering or protection if it is securely fastened to the construction; otherwise, it shall be protected in rigid metal conduit (RMC), intermediate metal conduit (IMC), rigid polyvinyl chloride conduit (PVC), reinforced thermosetting resin conduit (RTRC), electrical metallic tubing (EMT), or cable armor. Grounding electrode conductors smaller than 6 AWG shall be protected in RMC, IMC, PVC , RTRC, EMT, or cable armor.

(C) Continuous. Except as provided in 250.30(A)(5) and (A)(6), 250.30(B)(1), and 250.68(C), grounding electrode conductor(s) shall be installed in one continuous length without a splice or joint. If necessary, splices or connections shall be made as permitted in (1) through (4):

(1) Splicing of the wire-type grounding electrode conductor shall be permitted only by irreversible compression-type connectors listed as grounding and bonding equipment or by the exothermic welding process.
(2) Sections of busbars shall be permitted to be connected together to form a grounding electrode conductor.
(3) Bolted, riveted, or welded connections of structural metal frames of buildings or structures.
(4) Threaded, welded, brazed, soldered or bolted-flange connections of metal water piping.

(D) Service with Multiple Disconnecting Means Enclosures. If a service consists of more than a single enclosure as permitted in 230.71(A), grounding electrode connections shall be made in accordance with 250.64(D)(1), (D)(2), or (D)(3).

(1) Common Grounding Electrode Conductor and Taps. A common grounding electrode conductor and grounding electrode conductor taps shall be installed. The common grounding electrode conductor shall be sized in accordance with 250.66, based on the sum of the circular mil area of the largest ungrounded service-entrance conductor(s). If the service-entrance conductors connect directly to a service drop or service lateral, the common grounding electrode conductor shall be sized in accordance with Table 250.66, Note 1.

A grounding electrode conductor tap shall extend to the inside of each service disconnecting means enclosure. The grounding electrode conductor taps shall be sized in accordance with 250.66 for the largest service-entrance conductor serving the individual enclosure. The tap conductors shall be connected to the common grounding electrode conductor by one of the following methods in such a manner that the common grounding electrode conductor remains without a splice or joint:

(1) Exothermic welding.
(2) Connectors listed as grounding and bonding equipment.
(3) Connections to an aluminum or copper busbar not less than 6 mm × 50 mm (1/4 in. × 2 in.). The busbar shall be securely fastened and shall be installed in an accessible location. Connections shall be made by a listed connector or by the exothermic welding process. If aluminum busbars are used, the installation shall comply with 250.64(A).

(2) Individual Grounding Electrode Conductors. A grounding electrode conductor shall be connected between the grounded conductor in each service equipment disconnecting means enclosure and the grounding electrode system. Each grounding electrode conductor shall be sized in accordance with 250.66 based on the service-entrance conductor(s) supplying the individual service disconnecting means.

(3) Common Location. A grounding electrode conductor shall be connected to the grounded service conductor(s) in a wireway or other accessible enclosure on the supply side of the service disconnecting means. The connection shall be made with exothermic welding or a connector listed as grounding and bonding equipment. The grounding electrode conductor shall be sized in accordance with 250.66 based on the service-entrance conductor(s) at the common location where the connection is made.

(E) Enclosures for Grounding Electrode Conductors. Ferrous metal enclosures for grounding electrode conductors shall be electrically continuous from the point of attachment to cabinets or equipment to the grounding electrode and shall be securely fastened to the ground clamp or fitting. Nonferrous metal enclosures shall not be required to be electrically continuous. Ferrous metal enclosures that are not physically continuous from cabinets or equipment to the grounding electrode shall be made electrically continuous by bonding each end of the raceway or enclosure to the grounding electrode conductor. Bonding methods in compliance with 250.92(B) for installations at service equipment locations and with 250.92(B)(2) through (B)(4) for other than service equipment locations shall apply at each end and to all intervening ferrous raceways, boxes, and enclosures between the cabinets or equipment and the grounding electrode. The bonding jumper for a grounding electrode conductor raceway or cable armor shall be the same size as, or larger than, the enclosed grounding electrode conductor. If a raceway is used as protection for a grounding electrode conductor, the installation shall comply with the requirements of the appropriate raceway article.

(F) Installation to Electrode(s). Grounding electrode conductor(s) and bonding jumpers interconnecting grounding electrodes shall be installed in accordance with (1), (2), or (3). The grounding electrode conductor shall be sized for the largest grounding electrode conductor required among all the electrodes connected to it.

(1) The grounding electrode conductor shall be permitted to be run to any convenient grounding electrode available in the grounding electrode system where the other electrode(s), if any, is connected by bonding jumpers that are installed in accordance with 250.53(C).

(2) Grounding electrode conductor(s) shall be permitted to be run to one or more grounding electrode(s) individually.

(3) Bonding jumper(s) from grounding electrode(s) shall be permitted to be connected to an aluminum or copper busbar not less than 6 mm × 50 mm (1/4 in. × 2 in.). The busbar shall be securely fastened and shall be installed in an accessible location. Connections shall be made by a listed connector or by the exothermic welding process. The grounding electrode conductor shall be permitted to be run to the busbar. Where aluminum busbars are used, the installation shall comply with 250.64(A).

250.66 Size of Alternating-Current Grounding Electrode Conductor. The size of the grounding electrode conductor at the service, at each building or structure where supplied by a feeder(s) or branch circuit(s), or at a separately derived system of a grounded or ungrounded ac system shall not be less than given in Table 250.66, except as permitted in 250.66(A) through (C).

Table 250.66 Grounding Electrode Conductor for Alternating-Current Systems

Size of Largest Ungrounded Service-Entrance Conductor or Equivalent Area for Parallel Conductors[a] (AWG/kcmil)		Size of Grounding Electrode Conductor (AWG/kcmil)	
Copper	Aluminum or Copper Clad Aluminum	Copper	Aluminum or Copper-Clad Aluminum[b]
2 or smaller	1/0 or smaller	8	6
1 or 1/0	2/0 or 3/0	6	4
2/0 or 3/0	4/0 or 250	4	2
Over 3/0 through 350	Over 250 through 500	2	1/0
Over 350 through 600	Over 500 through 900	1/0	3/0
Over 600 through 1100	Over 900 through 1750	2/0	4/0
Over 1100	Over 1750	3/0	250

Notes:

1. Where multiple sets of service-entrance conductors are used as permitted in 230.40, Exception No. 2, the equivalent size of the largest service-entrance conductor shall be determined by the largest sum of the areas of the corresponding conductors of each set.

2. Where there are no service-entrance conductors, the grounding electrode conductor size shall be determined by the equivalent size of the largest service-entrance conductor required for the load to be served.

[a]This table also applies to the derived conductors of separately derived ac systems.

[b]See installation restrictions in 250.64(A).

(A) Connections to Rod, Pipe, or Plate Electrodes. Where the grounding electrode conductor is connected to rod, pipe, or plate electrodes as permitted in 250.52(A)(5) or (A)(7), that portion of the conductor that is the sole connection to the grounding electrode shall not be required to be larger than 6 AWG copper wire or 4 AWG aluminum wire.

(B) Connections to Concrete-Encased Electrodes.
Where the grounding electrode conductor is connected to
a concrete-encased electrode as permitted in 250.52(A)(3),
that portion of the conductor that is the sole connection to the
grounding electrode shall not be required to be larger than 4
AWG copper wire.

(C) Connections to Ground Rings. Where the grounding
electrode conductor is connected to a ground ring as permitted
in 250.52(A)(4), that portion of the conductor that is the sole
connection to the grounding electrode shall not be required to
be larger than the conductor used for the ground ring.

**250.68 Grounding Electrode Conductor and Bonding
Jumper Connection to Grounding Electrodes.** The connection of a grounding electrode conductor at the service,
at each building or structure where supplied by a feeder(s)
or branch circuit(s), or at a separately derived system and
associated bonding jumper(s) shall be made as specified
250.68(A) through (C).

(A) Accessibility. All mechanical elements used to terminate a grounding electrode conductor or bonding jumper to a
grounding electrode shall be accessible.

*Exception No. 1: An encased or buried connection to a con-
crete-encased, driven, or buried grounding electrode shall
not be required to be accessible.*

*Exception No. 2: Exothermic or irreversible compression
connections used at terminations, together with the mechanical means used to attach such terminations to fireproofed
structural metal whether or not the mechanical means is re-
versible, shall not be required to be accessible.*

(B) Effective Grounding Path. The connection of a grounding electrode conductor or bonding jumper to a grounding
electrode shall be made in a manner that will ensure an
effective grounding path. Where necessary to ensure the
grounding path for a metal piping system used as a grounding electrode, bonding shall be provided around insulated
joints and around any equipment likely to be disconnected
for repairs or replacement. Bonding jumpers shall be of

sufficient length to permit removal of such equipment while retaining the integrity of the grounding path.

(C) Metallic Water Pipe and Structural Metal. Grounding electrode conductors and bonding jumpers shall be permitted to be connected at the following locations and used to extend the connection to an electrode(s):

(1) Interior metal water piping located not more than 1.52 m (5 ft) from the point of entrance to the building shall be permitted to be used as a conductor to interconnect electrodes that are part of the grounding electrode system.

(2) The structural frame of a building that is directly connected to a grounding electrode as specified in 250.52(A) (2) or 250.68(C)(2)(a), (b), or (c) shall be permitted as a bonding conductor to interconnect electrodes that are part of the grounding electrode system, or as a grounding electrode conductor.

  a. By connecting the structural metal frame to the reinforcing bars of a concrete-encased electrode, as provided in 250.52(A)(3), or ground ring as provided in 250.52(A)(4)

  b. By bonding the structural metal frame to one or more of the grounding electrodes, as specified in 250.52(A) (5) or (A)(7), that comply with (2)

  c. By other approved means of establishing a connection to earth

250.70 Methods of Grounding and Bonding Conductor Connection to Electrodes. The grounding or bonding conductor shall be connected to the grounding electrode by exothermic welding, listed lugs, listed pressure connectors, listed clamps, or other listed means. Connections depending on solder shall not be used. Ground clamps shall be listed for the materials of the grounding electrode and the grounding electrode conductor and, where used on pipe, rod, or other buried electrodes, shall also be listed for direct soil burial or concrete encasement. Not more than one conductor shall be connected to the grounding electrode by a single clamp or fitting unless the clamp or fitting is listed for multiple conductors. One of the following methods shall be used:

(1) A pipe fitting, pipe plug, or other approved device screwed into a pipe or pipe fitting

(2) A listed bolted clamp of cast bronze or brass, or plain or malleable iron

(3) For indoor communications purposes only, a listed sheet metal strap-type ground clamp having a rigid metal base that seats on the electrode and having a strap of such material and dimensions that it is not likely to stretch during or after installation

(4) An equally substantial approved means

IV. ENCLOSURE, RACEWAY, AND SERVICE CABLE CONNECTIONS

250.80 Service Raceways and Enclosures. Metal enclosures and raceways for service conductors and equipment shall be connected to the grounded system conductor if the electrical system is grounded or to the grounding electrode conductor for electrical systems that are not grounded.

Exception: A metal elbow that is installed in an underground nonmetallic raceway and is isolated from possible contact by a minimum cover of 450 mm (18 in.) to any part of the elbow shall not be required to be connected to the grounded system conductor or grounding electrode conductor.

V. BONDING

250.90 General. Bonding shall be provided where necessary to ensure electrical continuity and the capacity to conduct safely any fault current likely to be imposed.

250.92 Services.

(A) Bonding of Equipment for Services. The normally non–current-carrying metal parts of equipment indicated in 250.92(A)(1) and (A)(2) shall be bonded together.

(1) All raceways, cable trays, cablebus framework, auxiliary gutters, or service cable armor or sheath that enclose, contain, or support service conductors, except as permitted in 250.80

(2) All enclosures containing service conductors, including meter fittings, boxes, or the like, interposed in the service raceway or armor

(B) Method of Bonding at the Service. Bonding jumpers meeting the requirements of this article shall be used around impaired connections, such as reducing washers or oversized, concentric, or eccentric knockouts. Standard locknuts or bushings shall not be the only means for the bonding required by this section but shall be permitted to be installed to make a mechanical connection of the raceway(s).

Electrical continuity at service equipment, service raceways, and service conductor enclosures shall be ensured by one of the following methods:

(1) Bonding equipment to the grounded service conductor in a manner provided in 250.8
(2) Connections utilizing threaded couplings or threaded hubs on enclosures if made up wrenchtight
(3) Threadless couplings and connectors if made up tight for metal raceways and metal-clad cables
(4) Other listed devices, such as bonding-type locknuts, bushings, or bushings with bonding jumpers

250.94 Bonding for Other Systems. An intersystem bonding termination for connecting intersystem bonding conductors required for other systems shall be provided external to enclosures at the service equipment or metering equipment enclosure and at the disconnecting means for any additional buildings or structures. The intersystem bonding termination shall comply with the following:

(1) Be accessible for connection and inspection.
(2) Consist of a set of terminals with the capacity for connection of not less than three intersystem bonding conductors.
(3) Not interfere with opening the enclosure for a service, building or structure disconnecting means, or metering equipment.
(4) At the service equipment, be securely mounted and electrically connected to an enclosure for the service equipment, to the meter enclosure, or to an exposed nonflexible metallic service raceway, or be mounted at

one of these enclosures and be connected to the enclosure or to the grounding electrode conductor with a minimum 6 AWG copper conductor

(5) At the disconnecting means for a building or structure, be securely mounted and electrically connected to the metallic enclosure for the building or structure disconnecting means, or be mounted at the disconnecting means and be connected to the metallic enclosure or to the grounding electrode conductor with a minimum 6 AWG copper conductor.

(6) The terminals shall be listed as grounding and bonding equipment.

Exception: In existing buildings or structures where any of the intersystem bonding and grounding electrode conductors required by 770.100(B)(2), 800.100(B)(2), 810.21(F)(2), 820.100(B)(2), and 830.100(B)(2) exist, installation of the intersystem bonding termination is not required. An accessible means external to enclosures for connecting intersystem bonding and grounding electrode conductors shall be permitted at the service equipment and at the disconnecting means for any additional buildings or structures by at least one of the following means:

(1) Exposed nonflexible metallic raceways

(2) An exposed grounding electrode conductor

(3) Approved means for the external connection of a copper or other corrosion-resistant bonding or grounding electrode conductor to the grounded raceway or equipment

250.96 Bonding Other Enclosures.

(A) General. Metal raceways, cable trays, cable armor, cable sheath, enclosures, frames, fittings, and other metal non–current-carrying parts that are to serve as equipment grounding conductors, with or without the use of supplementary equipment grounding conductors, shall be bonded where necessary to ensure electrical continuity and the capacity to conduct safely any fault current likely to be imposed on them. Any nonconductive paint, enamel, or similar coating shall be removed at threads, contact points, and contact surfaces or be connected by means of fittings designed so as to make such removal unnecessary.

250.102 Bonding Conductors and Jumpers.

(A) Material. Bonding jumpers shall be of copper or other corrosion-resistant material. A bonding jumper shall be a wire, bus, screw, or similar suitable conductor.

(B) Attachment. Bonding jumpers shall be attached in the manner specified by the applicable provisions of 250.8 for circuits and equipment and by 250.70 for grounding electrodes.

(C) Size—Supply-Side Bonding Jumper.

(1) Size for Supply Conductors in a Single Raceway or Cable. The supply-side bonding jumper shall not be smaller than the sizes shown in Table 250.66 for grounding electrode conductors. Where the ungrounded supply conductors are larger than 1100 kcmil copper or 1750 kcmil aluminum, the supply-side bonding jumper shall have an area not less than 12-1/2 percent of the area of the largest set of ungrounded supply conductors.

(2) Size for Parallel Conductor Installations. Where the ungrounded supply conductors are paralleled in two or more raceways or cables, and an individual supply-side bonding jumper is used for bonding these raceways or cables, the size of the supply-side bonding jumper for each raceway or cable shall be selected from Table 250.66 based on the size of the ungrounded supply conductors in each raceway or cable. A single supply-side bonding jumper installed for bonding two or more raceways or cables shall be sized in accordance with 250.102(C)(1).

(3) Different Materials. Where the ungrounded supply conductors and the supply-side bonding jumper are of different materials (copper or aluminum), the minimum size of the supply-side bonding jumper shall be based on the assumed use of ungrounded conductors of the same material as the supply-side bonding jumper and with an ampacity equivalent to that of the installed ungrounded supply conductors.

(D) Size—Equipment Bonding Jumper on Load Side of an Overcurrent Device. The equipment bonding jumper on the load side of an overcurrent device(s) shall be sized in accordance with 250.122.

A single common continuous equipment bonding jumper shall be permitted to connect two or more raceways or cables if the bonding jumper is sized in accordance with 250.122 for the largest overcurrent device supplying circuits therein.

(E) Installation. Bonding jumpers or conductors and equipment bonding jumpers shall be permitted to be installed inside or outside of a raceway or an enclosure.

(1) Inside a Raceway or an Enclosure. If installed inside a raceway, equipment bonding jumpers and bonding jumpers or conductors shall comply with the requirements of 250.119 and 250.148.

(2) Outside a Raceway or an Enclosure. If installed on the outside, the length of the bonding jumper or conductor or equipment bonding jumper shall not exceed 1.8 m (6 ft) and shall be routed with the raceway or enclosure.

Exception: An equipment bonding jumper or supply-side bonding jumper longer than 1.8 m (6 ft) shall be permitted at outside pole locations for the purpose of bonding or grounding isolated sections of metal raceways or elbows installed in exposed risers of metal conduit or other metal raceway, and for bonding grounding electrodes, and shall not be required to be routed with a raceway or enclosure.

(3) Protection. Bonding jumpers or conductors and equipment bonding jumpers shall be installed in accordance with 250.64(A) and (B).

250.104 Bonding of Piping Systems and Exposed Structural Steel.

(A) Metal Water Piping. The metal water piping system shall be bonded as required in (A)(1), (A)(2), or (A)(3) of this section. The bonding jumper(s) shall be installed in accordance with 250.64(A), (B), and (E). The points of attachment of the bonding jumper(s) shall be accessible.

(1) General. Metal water piping system(s) installed in or attached to a building or structure shall be bonded to the service equipment enclosure, the grounded conductor at the service, the grounding electrode conductor where of sufficient size, or to the one or more grounding electrodes used.

The bonding jumper(s) shall be sized in accordance with Table 250.66 except as permitted in 250.104(A)(2) and (A)(3).

(2) Buildings of Multiple Occupancy. In buildings of multiple occupancy where the metal water piping system(s) installed in or attached to a building or structure for the individual occupancies is metallically isolated from all other occupancies by use of nonmetallic water piping, the metal water piping system(s) for each occupancy shall be permitted to be bonded to the equipment grounding terminal of the panelboard or switchboard enclosure (other than service equipment) supplying that occupancy. The bonding jumper shall be sized in accordance with Table 250.122, based on the rating of the overcurrent protective device for the circuit supplying the occupancy.

(3) Multiple Buildings or Structures Supplied by a Feeder(s) or Branch Circuit(s). The metal water piping system(s) installed in or attached to a building or structure shall be bonded to the building or structure disconnecting means enclosure where located at the building or structure, to the equipment grounding conductor run with the supply conductors, or to the one or more grounding electrodes used. The bonding jumper(s) shall be sized in accordance with 250.66, based on the size of the feeder or branch circuit conductors that supply the building. The bonding jumper shall not be required to be larger than the largest ungrounded feeder or branch circuit conductor supplying the building.

(B) Other Metal Piping. If installed in, or attached to, a building or structure, a metal piping system(s), including gas piping, that is likely to become energized shall be bonded to the service equipment enclosure; the grounded conductor at the service; the grounding electrode conductor, if of sufficient size; or to one or more grounding electrodes used. The bonding conductor(s) or jumper(s) shall be sized in accordance with 250.122, using the rating of the circuit that is likely to energize the piping system(s). The equipment grounding conductor for the circuit that is likely to energize the piping shall be permitted to serve as the bonding means. The points of attachment of the bonding jumper(s) shall be accessible.

(C) Structural Metal. Exposed structural metal that is interconnected to form a metal building frame and is not intentionally grounded or bonded and is likely to become energized shall be bonded to the service equipment enclosure; the grounded conductor at the service; the disconnecting means for buildings or structures supplied by a feeder or branch circuit; the grounding electrode conductor, if of sufficient size; or to one or more grounding electrodes used. The bonding jumper(s) shall be sized in accordance with Table 250.66 and installed in accordance with 250.64(A), (B), and (E). The points of attachment of the bonding jumper(s) shall be accessible unless installed in compliance with 250.68(A), Exception No. 2.

VI. EQUIPMENT GROUNDING AND EQUIPMENT GROUNDING CONDUCTORS

250.110 Equipment Fastened in Place (Fixed) or Connected by Permanent Wiring Methods. Exposed, normally non–current-carrying metal parts of fixed equipment supplied by or enclosing conductors or components that are likely to become energized shall be connected to an equipment grounding conductor under any of the following conditions:

(1) Where within 2.5 m (8 ft) vertically or 1.5 m (5 ft) horizontally of ground or grounded metal objects and subject to contact by persons
(2) Where located in a wet or damp location and not isolated
(3) Where in electrical contact with metal

250.112 Specific Equipment Fastened in Place (Fixed) or Connected by Permanent Wiring Methods. Except as permitted in 250.112(F) and (I), exposed, normally non–current-carrying metal parts of equipment described in 250.112(A) through (K), and normally non–current-carrying metal parts of equipment and enclosures described in 250.112(L) and (M), shall be connected to an equipment grounding conductor, regardless of voltage.

(J) Luminaires. Luminaires as provided in Part V of Article 410.

(L) Motor-Operated Water Pumps. Motor-operated water pumps, including the submersible type.

(M) Metal Well Casings. Where a submersible pump is used in a metal well casing, the well casing shall be connected to the pump circuit equipment grounding conductor.

250.114 Equipment Connected by Cord and Plug. Under any of the conditions described in 250.114(1) through (4), exposed, normally non–current-carrying metal parts of cord-and-plug-connected equipment shall be connected to the equipment grounding conductor.

(3) In residential occupancies:

    a. Refrigerators, freezers, and air conditioners

    b. Clothes-washing, clothes-drying, dish-washing machines; ranges; kitchen waste disposers; information technology equipment; sump pumps and electrical aquarium equipment

    c. Hand-held motor-operated tools, stationary and fixed motor-operated tools, and light industrial motor-operated tools

    d. Motor-operated appliances of the following types: hedge clippers, lawn mowers, snow blowers, and wet scrubbers

    e. Portable handlamps

250.118 Types of Equipment Grounding Conductors. The equipment grounding conductor run with or enclosing the circuit conductors shall be one or more or a combination of the following:

(1) A copper, aluminum, or copper-clad aluminum conductor. This conductor shall be solid or stranded; insulated, covered, or bare; and in the form of a wire or a busbar of any shape.

(2) Rigid metal conduit.

(3) Intermediate metal conduit.

(4) Electrical metallic tubing.

(5) Listed flexible metal conduit meeting all the following conditions:

    a. The conduit is terminated in listed fittings.

    b. The circuit conductors contained in the conduit are protected by overcurrent devices rated at 20 amperes or less.

    c. The combined length of flexible metal conduit and flexible metallic tubing and liquidtight flexible metal conduit in the same ground-fault current path does not exceed 1.8 m (6 ft).

    d. If used to connect equipment where flexibility is necessary to minimize the transmission of vibration from equipment or to provide flexibility for equipment that requires movement after installation, an equipment grounding conductor shall be installed.

(6) Listed liquidtight flexible metal conduit meeting all the following conditions:

    a. The conduit is terminated in listed fittings.

    b. For metric designators 12 through 16 (trade sizes 3/8 through 1/2), the circuit conductors contained in the conduit are protected by overcurrent devices rated at 20 amperes or less.

    c. For metric designators 21 through 35 (trade sizes 3/4 through 1-1/4), the circuit conductors contained in the conduit are protected by overcurrent devices rated not more than 60 amperes and there is no flexible metal conduit, flexible metallic tubing, or liquidtight flexible metal conduit in trade sizes metric designators 12 through 16 (trade sizes 3/8 through 1/2) in the ground-fault current path.

    d. The combined length of flexible metal conduit and flexible metallic tubing and liquidtight flexible metal conduit in the same ground-fault current path does not exceed 1.8 m (6 ft).

    e. If used to connect equipment where flexibility is necessary to minimize the transmission of vibration from equipment or to provide flexibility for equipment that requires movement after installation, an equipment grounding conductor shall be installed.

(7) Flexible metallic tubing where the tubing is terminated in listed fittings and meeting the following conditions:

a. The circuit conductors contained in the tubing are protected by overcurrent devices rated at 20 amperes or less.

b. The combined length of flexible metal conduit and flexible metallic tubing and liquidtight flexible metal conduit in the same ground-fault current path does not exceed 1.8 m (6 ft).

(8) Armor of Type AC cable as provided in 320.108.

(9) The copper sheath of mineral-insulated, metal-sheathed cable.

(10) Type MC cable that provides an effective ground-fault current path in accordance with one or more of the following:

a. It contains an insulated or uninsulated equipment grounding conductor in compliance with 250.118(1)

b. The combined metallic sheath and uninsulated equipment grounding/bonding conductor of inter-locked metal tape–type MC cable that is listed and identified as an equipment grounding conductor

c. The metallic sheath or the combined metallic sheath and equipment grounding conductors of the smooth or corrugated tube-type MC cable that is listed and identified as an equipment grounding conductor

250.119 Identification of Equipment Grounding Conductors. Unless required elsewhere in this *Code,* equipment grounding conductors shall be permitted to be bare, covered, or insulated. Individually covered or insulated equipment grounding conductors shall have a continuous outer finish that is either green or green with one or more yellow stripes except as permitted in this section. Conductors with insulation or individual covering that is green, green with one or more yellow stripes, or otherwise identified as permitted by this section shall not be used for ungrounded or grounded circuit conductors.

Exception: Power-limited Class 2 or Class 3 cables, power-limited fire alarm cables, or communications cables containing only circuits operating at less than 50 volts where connected to equipment not required to be grounded in accordance with 250.112(I) shall be permitted to use a conduc-

tor with green insulation or green with one or more yellow stripes for other than equipment grounding purposes.

(A) Conductors Larger Than 6 AWG. Equipment grounding conductors larger than 6 AWG shall comply with 250.119(A) (1) and (A)(2).

(1) An insulated or covered conductor larger than 6 AWG shall be permitted, at the time of installation, to be permanently identified as an equipment grounding conductor at each end and at every point where the conductor is accessible.

Exception: Conductors larger than 6 AWG shall not be required to be marked in conduit bodies that contain no splices or unused hubs.

(2) Identification shall encircle the conductor and shall be accomplished by one of the following:

   a. Stripping the insulation or covering from the entire exposed length
   b. Coloring the insulation or covering green at the termination
   c. Marking the insulation or covering with green tape or green adhesive labels at the termination

250.120 Equipment Grounding Conductor Installation. An equipment grounding conductor shall be installed in accordance with 250.120(A), (B), and (C).

(A) Raceway, Cable Trays, Cable Armor, Cablebus, or Cable Sheaths. Where it consists of a raceway, cable tray, cable armor, cablebus framework, or cable sheath or where it is a wire within a raceway or cable, it shall be installed in accordance with the applicable provisions in this *Code* using fittings for joints and terminations approved for use with the type raceway or cable used. All connections, joints, and fittings shall be made tight using suitable tools.

(B) Aluminum and Copper-Clad Aluminum Conductors. Equipment grounding conductors of bare or insulated aluminum or copper-clad aluminum shall be permitted. Bare conductors shall not come in direct contact with masonry or the earth or where subject to corrosive conditions. Aluminum or copper-clad aluminum conductors shall not be terminated within 450 mm (18 in.) of the earth.

(C) Equipment Grounding Conductors Smaller Than 6 AWG. Where not routed with circuit conductors as permitted in 250.130(C) and 250.134(B) Exception No. 2, equipment grounding conductors smaller than 6 AWG shall be protected from physical damage by an identified raceway or cable armor unless installed within hollow spaces of the framing members of buildings or structures and where not subject to physical damage.

250.121 Use of Equipment Grounding Conductors. An equipment grounding conductor shall not be used as a grounding electrode conductor.

250.122 Size of Equipment Grounding Conductors.

(A) General. Copper, aluminum, or copper-clad aluminum equipment grounding conductors of the wire type shall not be smaller than shown in   250.122, but in no case shall they be required to be larger than the circuit conductors supplying the equipment. Where a cable tray, a raceway, or a cable armor or sheath is used as the equipment grounding conductor, as provided in 250.118 and 250.134(A), it shall comply with 250.4(A)(5) or (B)(4).

Equipment grounding conductors shall be permitted to be sectioned within a multiconductor cable, provided the combined circular mil area complies with Table 250.122.

(B) Increased in Size. Where ungrounded conductors are increased in size, equipment grounding conductors, where installed, shall be increased in size proportionately according to the circular mil area of the ungrounded conductors.

(C) Multiple Circuits. Where a single equipment grounding conductor is run with multiple circuits in the same raceway, cable, or cable tray, it shall be sized for the largest overcurrent device protecting conductors in the raceway, cable, or cable tray. Equipment grounding conductors installed in cable trays shall meet the minimum requirements of 392.10(B)(1)(c).

(F) Conductors in Parallel. Where conductors are installed in parallel in multiple raceways or cables as permitted

Table 250.122 Minimum Size Equipment Grounding Conductors for Grounding Raceway and Equipment

Rating or Setting of Automatic Overcurrent Device in Circuit Ahead of Equipment, Conduit, etc., Not Exceeding (Amperes)	Size (AWG or kcmil)	
	Copper	Aluminum or Copper-Clad Aluminum*
15	14	12
20	12	10
60	10	8
100	8	6
200	6	4
300	4	2
400	3	1
500	2	1/0
600	1	2/0
800	1/0	3/0
1000	2/0	4/0
1200	3/0	250
1600	4/0	350
2000	250	400
2500	350	600
3000	400	600
4000	500	750
5000	700	1200
6000	800	1200

Note: Where necessary to comply with 250.4(A)(5) or (B)(4), the equipment grounding conductor shall be sized larger than given in this table. *See installation restrictions in 250.120.

in 310.10(H), the equipment grounding conductors, where used, shall be installed in parallel in each raceway or cable. Where conductors are installed in parallel in the same raceway, cable, or cable tray as permitted in 310.10(H), a single equipment grounding conductor shall be permitted. Equipment grounding conductors installed in cable tray shall meet the minimum requirements of 392.10(B)(1)(c).

Each equipment grounding conductor shall be sized in compliance with 250.122.

(G) Feeder Taps. Equipment grounding conductors run with feeder taps shall not be smaller than shown in Table 250.122 based on the rating of the overcurrent device ahead of the feeder but shall not be required to be larger than the tap conductors.

250.126 Identification of Wiring Device Terminals. The terminal for the connection of the equipment grounding conductor shall be identified by one of the following:

(1) A green, not readily removable terminal screw with a hexagonal head.

(2) A green, hexagonal, not readily removable terminal nut.

(3) A green pressure wire connector. If the terminal for the grounding conductor is not visible, the conductor entrance hole shall be marked with the word *green* or *ground,* the letters *G* or *GR,* a grounding symbol, or otherwise identified by a distinctive green color. If the terminal for the equipment grounding conductor is readily removable, the area adjacent to the terminal shall be similarly marked.

VII. METHODS OF EQUIPMENT GROUNDING

250.130 Equipment Grounding Conductor Connections. Equipment grounding conductor connections at the source of separately derived systems shall be made in accordance with 250.30(A)(1). Equipment grounding conductor connections at service equipment shall be made as indicated in 250.130(A) or (B). For replacement of non–grounding-type receptacles with grounding-type receptacles and for branch-circuit extensions only in existing installations that do not have an equipment grounding conductor in the branch circuit, connections shall be permitted as indicated in 250.130(C).

(A) For Grounded Systems. The connection shall be made by bonding the equipment grounding conductor to the grounded service conductor and the grounding electrode conductor.

(B) For Ungrounded Systems. The connection shall be made by bonding the equipment grounding conductor to the grounding electrode conductor.

(C) Nongrounding Receptacle Replacement or Branch Circuit Extensions. The equipment grounding conductor of a grounding-type receptacle or a branch-circuit extension shall be permitted to be connected to any of the following:

(1) Any accessible point on the grounding electrode system as described in 250.50
(2) Any accessible point on the grounding electrode conductor
(3) The equipment grounding terminal bar within the enclosure where the branch circuit for .the receptacle or branch circuit originates
(4) For grounded systems, the grounded service conductor within the service equipment enclosure
(5) For ungrounded systems, the grounding terminal bar within the service equipment enclosure

250.132 Short Sections of Raceway. Isolated sections of metal raceway or cable armor, where required to be grounded, shall be connected to an equipment grounding conductor in accordance with 250.134.

250.134 Equipment Fastened in Place or Connected by Permanent Wiring Methods (Fixed)—Grounding. Unless grounded by connection to the grounded circuit conductor as permitted by 250.32, 250.140, and 250.142, non–current-carrying metal parts of equipment, raceways, and other enclosures, if grounded, shall be connected to an equipment grounding conductor by one of the methods specified in 250.134(A) or (B).

(A) Equipment Grounding Conductor Types. By connecting to any of the equipment grounding conductors permitted by 250.118.

(B) With Circuit Conductors. By connecting to an equipment grounding conductor contained within the same raceway, cable, or otherwise run with the circuit conductors.

250.138 Cord-and-Plug-Connected Equipment. Non–current-carrying metal parts of cord-and-plug-connected equipment, if grounded, shall be connected to an equipment grounding conductor by one of the methods in 250.138(A) or (B).

(A) By Means of an Equipment Grounding Conductor. By means of an equipment grounding conductor run with the power supply conductors in a cable assembly or flexible cord properly terminated in a grounding-type attachment plug with one fixed grounding contact.

(B) By Means of a Separate Flexible Wire or Strap. By means of a separate flexible wire or strap, insulated or bare, connected to an equipment grounding conductor, and protected as well as practicable against physical damage, where part of equipment.

250.140 Frames of Ranges and Clothes Dryers. Frames of electric ranges, wall-mounted ovens, counter-mounted cooking units, clothes dryers, and outlet or junction boxes that are part of the circuit for these appliances shall be connected to the equipment grounding conductor in the manner specified by 250.134 or 250.138.

Exception: For existing branch-circuit installations only where an equipment grounding conductor is not present in the outlet or junction box, the frames of electric ranges, wall-mounted ovens, counter-mounted cooking units, clothes dryers, and outlet or junction boxes that are part of the circuit for these appliances shall be permitted to be connected to the grounded circuit conductor if all the following conditions are met.

(1) The supply circuit is 120/240-volt, single-phase, 3-wire; or 208Y/120-volt derived from a 3-phase, 4-wire, wye-connected system.

(2) The grounded conductor is not smaller than 10 AWG copper or 8 AWG aluminum.

(3) The grounded conductor is insulated, or the grounded conductor is uninsulated and part of a Type SE service-entrance cable and the branch circuit originates at the service equipment.

(4) Grounding contacts of receptacles furnished as part of the equipment are bonded to the equipment.

250.142 Use of Grounded Circuit Conductor for Grounding Equipment.

(A) Supply-Side Equipment. A grounded circuit conductor shall be permitted to ground non–current-carrying metal parts of equipment, raceways, and other enclosures at any of the following locations:

(1) On the supply side or within the enclosure of the ac service-disconnecting means
(2) On the supply side or within the enclosure of the main disconnecting means for separate buildings as provided in 250.32(B)

(B) Load-Side Equipment. Except as permitted in 250.30(A)(1) and 250.32(B) exception, a grounded circuit conductor shall not be used for grounding non–current-carrying metal parts of equipment on the load side of the service disconnecting means or on the load side of a separately derived system disconnecting means or the overcurrent devices for a separately derived system not having a main disconnecting means.

Exception No. 1: The frames of ranges, wall-mounted ovens, counter-mounted cooking units, and clothes dryers under the conditions permitted for existing installations by 250.140 shall be permitted to be connected to the grounded circuit conductor.

Exception No. 2: It shall be permissible to ground meter enclosures by connection to the grounded circuit conductor on the load side of the service disconnect where all of the following conditions apply:

(1) No service ground-fault protection is installed.
(2) All meter enclosures are located immediately adjacent to the service disconnecting means.
(3) The size of the grounded circuit conductor is not smaller than the size specified in Table 250.122 for equipment grounding conductors.

250.146 Connecting Receptacle Grounding Terminal to Box. An equipment bonding jumper shall be used to connect the grounding terminal of a grounding-type receptacle to a grounded box unless grounded as in 250.146(A) through (D). The equipment bonding jumper shall be sized in accordance with Table 250.122 based on the rating of the overcurrent device protecting the circuit conductors.

(A) Surface-Mounted Box. Where the box is mounted on the surface, direct metal-to-metal contact between the device yoke and the box or a contact yoke or device that complies with 250.146(B) shall be permitted to ground the receptacle to the box. At least one of the insulating washers shall be removed from receptacles that do not have a contact yoke or device that complies with 250.146(B) to ensure direct metal-to-metal contact. This provision shall not apply to cover-mounted receptacles unless the box and cover combination are listed as providing satisfactory ground continuity between the box and the receptacle. A listed exposed work cover shall be permitted to be the grounding and bonding means when (1) the device is attached to the cover with at least two fasteners that are permanent (such as a rivet) or have a thread locking or screw or nut locking means and (2) when the cover mounting holes are located on a flat non-raised portion of the cover.

(B) Contact Devices or Yokes. Contact devices or yokes designed and listed as self-grounding shall be permitted in conjunction with the supporting screws to establish the grounding circuit between the device yoke and flush-type boxes.

(C) Floor Boxes. Floor boxes designed for and listed as providing satisfactory ground continuity between the box and the device shall be permitted.

(D) Isolated Receptacles. Where installed for the reduction of electrical noise (electromagnetic interference) on the grounding circuit, a receptacle in which the grounding terminal is purposely insulated from the receptacle mounting means shall be permitted. The receptacle grounding terminal shall be connected to an insulated equipment grounding conductor run with the circuit conductors. This equipment grounding conductor shall be permitted to pass through one or more panelboards without a connection to the panelboard grounding terminal bar as permitted in 408.40, Exception, so as to terminate within the same building or structure directly at an equipment grounding conductor terminal of the applicable derived system or service. Where installed in accordance with the provisions of this section, this equipment grounding conductor shall also be permitted to pass through boxes, wireways, or other enclosures without being connected to such enclosures.

> Informational Note: Use of an isolated equipment grounding conductor does not relieve the requirement for grounding the raceway system and outlet box.

250.148 Continuity and Attachment of Equipment Grounding Conductors to Boxes. Where circuit conductors are spliced within a box, or terminated on equipment within or supported by a box, any equipment grounding conductor(s) associated with those circuit conductors shall be connected within the box or to the box with devices suitable for the use in accordance with 250.148(A) through (E).

Exception: The equipment grounding conductor permitted in 250.146(D) shall not be required to be connected to the other equipment grounding conductors or to the box.

(A) Connections. Connections and splices shall be made in accordance with 110.14(B) except that insulation shall not be required.

(B) Grounding Continuity. The arrangement of grounding connections shall be such that the disconnection or the removal of a receptacle, luminaire, or other device fed from the box does not interfere with or interrupt the grounding continuity.

(C) Metal Boxes. A connection shall be made between the one or more equipment grounding conductors and a metal box by means of a grounding screw that shall be used for no other purpose, equipment listed for grounding, or a listed grounding device.

(D) Nonmetallic Boxes. One or more equipment grounding conductors brought into a nonmetallic outlet box shall be arranged such that a connection can be made to any fitting or device in that box requiring grounding.

(E) Solder. Connections depending solely on solder shall not be used.

CHAPTER 12

BOXES AND ENCLOSURES

INTRODUCTION

This chapter contains provisions from Article 314—Outlet, Device, Pull and Junction Boxes; Conduit Bodies; and Fittings; and Handhole Enclosures. Article 314 covers the installation and use of all enclosures (and conduit bodies) used as outlet, device, pull, or junction boxes, depending on their use. Included also are installation requirements for fittings used to join raceways and to connect raceways (and cables) to boxes and conduit bodies. This article's provisions range from selecting the size and type of box to installing the cover or canopy.

Provisions for boxes enclosing 18 through 6 AWG conductors are located in 314.16. Boxes (outlet, device, and junction) covered by 314.16 are calculated from the sizes and numbers of conductors. Requirements for boxes containing larger conductors (4 AWG and larger) are in 314.28. Pull and junction boxes covered by 314.28 are calculated from the sizes and numbers of raceways.

ARTICLE 314

Outlet, Device, Pull, and Junction Boxes; Conduit Bodies; Fittings; and Handhole Enclosures

I. SCOPE AND GENERAL

314.1 Scope. This article covers the installation and use of all boxes and conduit bodies used as outlet, device, junction, or pull boxes, depending on their use, and handhole enclosures. Cast, sheet metal, nonmetallic, and other boxes such as FS, FD, and larger boxes are not classified as conduit bodies. This article also includes installation requirements for fittings used to join raceways and to connect raceways and cables to boxes and conduit bodies.

314.2 Round Boxes. Round boxes shall not be used where conduits or connectors requiring the use of locknuts or bushings are to be connected to the side of the box.

314.3 Nonmetallic Boxes. Nonmetallic boxes shall be permitted only with open wiring on insulators, concealed knob-and-tube wiring, cabled wiring methods with entirely nonmetallic sheaths, flexible cords, and nonmetallic raceways.

Exception No. 1: Where internal bonding means are provided between all entries, nonmetallic boxes shall be permitted to be used with metal raceways or metal-armored cables.

Exception No. 2: Where integral bonding means with a provision for attaching an equipment bonding jumper inside the box are provided between all threaded entries in nonmetallic boxes listed for the purpose, nonmetallic boxes shall be permitted to be used with metal raceways or metal-armored cables.

314.4 Metal Boxes. Metal boxes shall be grounded and bonded in accordance with Parts I, IV, V, VI, VII, and X of Article 250 as applicable, except as permitted in 250.112(I).

II. INSTALLATION

314.15 Damp or Wet Locations. In damp or wet locations, boxes, conduit bodies, and fittings shall be placed or equipped so as to prevent moisture from entering or accumulating within the box, conduit body, or fitting. Boxes, conduit bodies, and fittings installed in wet locations shall be listed for use in wet locations.

314.16 Number of Conductors in Outlet, Device, and Junction Boxes, and Conduit Bodies. Boxes and conduit bodies shall be of sufficient size to provide free space for all enclosed conductors. In no case shall the volume of the box, as calculated in 314.16(A), be less than the fill calculation as calculated in 314.16(B). The minimum volume for conduit bodies shall be as calculated in 314.16(C).

The provisions of this section shall not apply to terminal housings supplied with motors or generators.

Boxes and conduit bodies enclosing conductors 4 AWG or larger shall also comply with the provisions of 314.28.

(A) Box Volume Calculations. The volume of a wiring enclosure (box) shall be the total volume of the assembled sections and, where used, the space provided by plaster rings, domed covers, extension rings, and so forth, that are marked with their volume or are made from boxes the dimensions of which are listed in Table 314.16(A).

(1) Standard Boxes. The volumes of standard boxes that are not marked with their volume shall be as given in Table 314.16(A).

(2) Other Boxes. Boxes 1650 cm³ (100 in.³) or less, other than those described in Table 314.16(A), and nonmetallic boxes shall be durably and legibly marked by the manufacturer with their volume. Boxes described in Table 314.16(A) that have a volume larger than is designated in the table shall be permitted to have their volume marked as required by this section.

(B) Box Fill Calculations. The volumes in paragraphs 314.16(B)(1) through (B)(5), as applicable, shall be added together. No allowance shall be required for small fittings such as locknuts and bushings.

(1) Conductor Fill. Each conductor that originates outside the box and terminates or is spliced within the box shall be counted once, and each conductor that passes through the

Table 314.16(A) Metal Boxes

Box Trade Size		Minimum Volume		Maximum Number of Conductors* (arranged by AWG size)							
mm	in.	cm³	in.³	18	16	14	12	10	8	6
100 × 32	(4 × 1-1/4) round/octagonal	205	12.5	8	7	6	5	5	5	2
100 × 38	(4 × 1-1/2) round/octagonal	254	15.5	10	8	7	6	6	5	3
100 × 54	(4 × 2-1/8) round/octagonal	353	21.5	14	12	10	9	8	7	4
100 × 32	(4 × 1-1/4) square	295	18.0	12	10	9	8	7	6	3
100 × 38	(4 × 1-1/2) square	344	21.0	14	12	10	9	8	7	4
100 × 54	(4 × 2-1/8) square	497	30.3	20	17	15	13	12	10	6
120 × 32	(4-11/16 × 1-1/4) square	418	25.5	17	14	12	11	10	8	5
120 × 38	(4-11/16 × 1-1/2) square	484	29.5	19	16	14	13	11	9	5
120 × 54	(4-11/16 × 2-1/8) square	689	42.0	28	24	21	18	16	14	8
75 × 50 × 38	(3 × 2 × 1-1/2) device	123	7.5	5	4	3	3	3	2	1
75 × 50 × 50	(3 × 2 × 2) device	164	10.0	6	5	5	4	4	3	2
75 × 50 × 57	(3 × 2 × 2-1/4) device	172	10.5	7	6	5	4	4	3	2

75 × 50 × 65	(3 × 2 × 2-1/2)	device	205	12.5	8	7	6	5	5	4	2
75 × 50 × 70	(3 × 2 × 2-3/4)	device	230	14.0	9	8	7	6	5	4	2
75 × 50 × 90	(3 × 2 × 3-1/2)	device	295	18.0	12	10	9	8	7	6	3
100 × 54 × 38	(4 × 2-1/8 × 1-1/2)	device	169	10.3	6	5	5	4	4	3	2
100 × 54 × 48	(4 × 2-1/8 × 1-7/8)	device	213	13.0	8	7	6	5	5	4	2
100 × 54 × 54	(4 × 2-1/8 × 2-1/8)	device	238	14.5	9	8	7	6	5	4	2
95 × 50 × 65	(3-3/4 × 2 × 2-1/2)	masonry box/gang	230	14.0	9	8	7	6	5	4	2
95 × 50 × 90	(3-3/4 × 2 × 3-1/2)	masonry box/gang	344	21.0	14	12	10	9	8	7	4
min. 44.5 depth	FS—single cover/gang (1-3/4)		221	13.5	9	7	6	6	5	4	2
min. 60.3 depth	FD—single cover/gang (2-3/8)		295	18.0	12	10	8	8	7	6	3
min. 44.5 depth	FS—multiple cover/gang (1-3/4)		295	18.0	12	10	9	8	7	6	3
min. 60.3 depth	FD—multiple cover/gang (2-3/8)		395	24.0	16	13	12	10	9	8	4

*Where no volume allowances are required by 314.16(B)(2) through (B)(5).

Table 314.16(B) Volume Allowance Required per Conductor

Size of Conductor (AWG)	Free Space Within Box for Each Conductor	
	cm³	in.³
18	24.6	1.50
16	28.7	1.75
14	32.8	2.00
12	36.9	2.25
10	41.0	2.50
8	49.2	3.00
6	81.9	5.00

box without splice or termination shall be counted once. Each loop or coil of unbroken conductor not less than twice the minimum length required for free conductors in 300.14 shall be counted twice. The conductor fill shall be calculated using Table 314.16(B). A conductor, no part of which leaves the box, shall not be counted.

(2) Clamp Fill. Where one or more internal cable clamps, whether factory or field supplied, are present in the box, a single volume allowance in accordance with Table 314.16(B) shall be made based on the largest conductor present in the box. No allowance shall be required for a cable connector with its clamping mechanism outside the box.

(3) Support Fittings Fill. Where one or more luminaire studs or hickeys are present in the box, a single volume allowance in accordance with Table 314.16(B) shall be made for each type of fitting based on the largest conductor present in the box.

(4) Device or Equipment Fill. For each yoke or strap containing one or more devices or equipment, a double volume allowance in accordance with Table 314.16(B) shall be made for each yoke or strap based on the largest conductor connected to a device(s) or equipment supported by that yoke or strap. A device or utilization equipment wider than a single 50 mm (2 in.) device box as described in Table 314.16(A) shall have double volume allowances provided for each gang required for mounting.

(5) Equipment Grounding Conductor Fill. Where one or more equipment grounding conductors or equipment bonding jumpers enter a box, a single volume allowance in accordance with Table 314.16(B) shall be made based on the largest equipment grounding conductor or equipment bonding jumper present in the box. Where an additional set of equipment grounding conductors, as permitted by 250.146(D), is present in the box, an additional volume allowance shall be made based on the largest equipment grounding conductor in the additional set.

(C) Conduit Bodies.

(1) General. Conduit bodies enclosing 6 AWG conductors or smaller, other than short-radius conduit bodies as described in 314.16(C)(2), shall have a cross-sectional area not less than twice the cross-sectional area of the largest conduit or tubing to which they can be attached. The maximum number of conductors permitted shall be the maximum number permitted by Table 1 of Chapter 9 for the conduit or tubing to which it is attached.

(2) With Splices, Taps, or Devices. Only those conduit bodies that are durably and legibly marked by the manufacturer with their volume shall be permitted to contain splices, taps, or devices. The maximum number of conductors shall be calculated in accordance with 314.16(B). Conduit bodies shall be supported in a rigid and secure manner.

(3) Short Radius Conduit Bodies. Conduit bodies such as capped elbows and service-entrance elbows that enclose conductors 6 AWG or smaller, and are only intended to enable the installation of the raceway and the contained conductors, shall not contain splices, taps, or devices and shall be of sufficient size to provide free space for all conductors enclosed in the conduit body.

314.17 Conductors Entering Boxes, Conduit Bodies, or Fittings. Conductors entering boxes, conduit bodies, or fittings shall be protected from abrasion and shall comply with 314.17(A) through (D).

(A) Openings to Be Closed. Openings through which conductors enter shall be adequately closed.

(B) Metal Boxes and Conduit Bodies. Where metal boxes or conduit bodies are installed with messenger-supported wiring, open wiring on insulators, or concealed knob-and-tube wiring, conductors shall enter through insulating bushings or, in dry locations, through flexible tubing extending from the last insulating support to not less than 6 mm (1/4 in.) inside the box and beyond any cable clamps. Except as provided in 300.15(C), the wiring shall be firmly secured to the box or conduit body. Where raceway or cable is installed with metal boxes or conduit bodies, the raceway or cable shall be secured to such boxes and conduit bodies.

(C) Nonmetallic Boxes and Conduit Bodies. Nonmetallic boxes and conduit bodies shall be suitable for the lowest temperature-rated conductor entering the box. Where nonmetallic boxes and conduit bodies are used with messenger-supported wiring, open wiring on insulators, or concealed knob-and-tube wiring, the conductors shall enter the box through individual holes. Where flexible tubing is used to enclose the conductors, the tubing shall extend from the last insulating support to not less than 6 mm (1/4 in.) inside the box and beyond any cable clamp. Where nonmetallic-sheathed cable or multiconductor Type UF cable is used, the sheath shall extend not less than 6 mm (1/4 in.) inside the box and beyond any cable clamp. In all instances, all permitted wiring methods shall be secured to the boxes.

Exception: Where nonmetallic-sheathed cable or multiconductor Type UF cable is used with single gang boxes not larger than a nominal size 57 mm × 100 mm (2-1/4 in. × 4 in.) mounted in walls or ceilings, and where the cable is fastened within 200 mm (8 in.) of the box measured along the sheath and where the sheath extends through a cable knockout not less than 6 mm (1/4 in.), securing the cable to the box shall not be required. Multiple cable entries shall be permitted in a single cable knockout opening.

(D) Conductors 4 AWG or Larger. Installation shall comply with 300.4(G).

314.19 Boxes Enclosing Flush Devices. Boxes used to enclose flush devices shall be of such design that the devices will be completely enclosed on back and sides and substantial support for the devices will be provided. Screws for supporting the box shall not be used in attachment of the device contained therein.

314.20 In Wall or Ceiling. In walls or ceilings with a surface of concrete, tile, gypsum, plaster, or other noncombustible material, boxes employing a flush-type cover or faceplate shall be installed so that the front edge of the box, plaster ring, extension ring, or listed extender will not be set back of the finished surface more than 6 mm (1/4 in.).

In walls and ceilings constructed of wood or other combustible surface material, boxes, plaster rings, extension rings, or listed extenders shall be flush with the finished surface or project therefrom.

314.21 Repairing Noncombustible Surfaces. Noncombustible surfaces that are broken or incomplete around boxes employing a flush-type cover or faceplate shall be repaired so there will be no gaps or open spaces greater than 3 mm (1/8 in.) at the edge of the box.

314.22 Surface Extensions. Surface extensions shall be made by mounting and mechanically securing an extension ring over the box. Equipment grounding shall be in accordance with Part VI of Article 250.

Exception: A surface extension shall be permitted to be made from the cover of a box where the cover is designed so it is unlikely to fall off or be removed if its securing means becomes loose. The wiring method shall be flexible for a length sufficient to permit removal of the cover and provide access to the box interior, and arranged so that any grounding continuity is independent of the connection between the box and cover.

314.23 Supports. Enclosures within the scope of this article shall be supported in accordance with one or more of the provisions in 314.23(A) through (H).

(A) Surface Mounting. An enclosure mounted on a building or other surface shall be rigidly and securely fastened in place. If the surface does not provide rigid and secure support, additional support in accordance with other provisions of this section shall be provided.

(B) Structural Mounting. An enclosure supported from a structural member of a building or from grade shall be rigidly supported either directly or by using a metal, polymeric, or wood brace.

(1) Nails and Screws. Nails and screws, where used as a fastening means, shall be attached by using brackets on the outside of the enclosure, or they shall pass through the interior within 6 mm (1/4 in.) of the back or ends of the enclosure. Screws shall not be permitted to pass through the box unless exposed threads in the box are protected using approved means to avoid abrasion of conductor insulation.

(2) Braces. Metal braces shall be protected against corrosion and formed from metal that is not less than 0.51 mm (0.020 in.) thick uncoated. Wood braces shall have a cross section not less than nominal 25 mm × 50 mm (1 in. × 2 in.). Wood braces in wet locations shall be treated for the conditions. Polymeric braces shall be identified as being suitable for the use.

(C) Mounting in Finished Surfaces. An enclosure mounted in a finished surface shall be rigidly secured thereto by clamps, anchors, or fittings identified for the application.

(D) Suspended Ceilings. An enclosure mounted to structural or supporting elements of a suspended ceiling shall be not more than 1650 cm³ (100 in.³) in size and shall be securely fastened in place in accordance with either (D)(1) or (D)(2).

(1) Framing Members. An enclosure shall be fastened to the framing members by mechanical means such as bolts, screws, or rivets, or by the use of clips or other securing means identified for use with the type of ceiling framing member(s) and enclosure(s) employed. The framing members shall be adequately supported and securely fastened to each other and to the building structure.

(2) Support Wires. The installation shall comply with the provisions of 300.11(A). The enclosure shall be secured, using methods 1identified for the purpose, to ceiling support wire(s), including any additional support wire(s) installed for that purpose. Support wire(s) used for enclosure support shall be fastened at each end so as to be taut within the ceiling cavity.

(E) Raceway Supported Enclosure, Without Devices, Luminaires, or Lampholders. An enclosure that does not contain a device(s) other than splicing devices or support a luminaire(s), lampholder, or other equipment and is supported by entering raceways shall not exceed 1650 cm³ (100 in.³) in size. It shall have threaded entries or have hubs identified for the purpose. It shall be supported by two or more conduits threaded wrenchtight into the enclosure or hubs. Each conduit shall be secured within 900 mm (3 ft) of the enclosure, or within 450 mm (18 in.) of the enclosure if all conduit entries are on the same side.

Exception: The following wiring methods shall be permitted to support a conduit body of any size, including a conduit body constructed with only one conduit entry, if the trade size of the conduit body is not larger than the largest trade size of the conduit or tubing:

(1) Intermediate metal conduit, Type IMC

(2) Rigid metal conduit, Type RMC

(3) Rigid polyvinyl chloride conduit, Type PVC

(4) Reinforced thermosetting resin conduit, Type RTRC

(5) Electrical metallic tubing, Type EMT

(F) Raceway-Supported Enclosures, with Devices, Luminaires, or Lampholders. An enclosure that contains a device(s), other than splicing devices, or supports a luminaire(s), lampholder, or other equipment and is supported by entering raceways shall not exceed 1650 cm³ (100 in.³) in size. It shall have threaded entries or have hubs identified for the purpose. It shall be supported by two or more conduits threaded wrenchtight into the enclosure or hubs. Each conduit shall be secured within 450 mm (18 in.) of the enclosure.

Exception No. 1: Rigid metal or intermediate metal conduit shall be permitted to support a conduit body of any size, including a conduit body constructed with only one conduit entry, provided the trade size of the conduit body is not larger than the largest trade size of the conduit.

(G) Enclosures in Concrete or Masonry. An enclosure supported by embedment shall be identified as suitably protected from corrosion and securely embedded in concrete or masonry.

(H) Pendant Boxes. An enclosure supported by a pendant shall comply with 314.23(H)(1) or (H)(2).

(1) Flexible Cord. A box shall be supported from a multiconductor cord or cable in an approved manner that protects the conductors against strain, such as a strain-relief connector threaded into a box with a hub.

(2) Conduit. A box supporting lampholders or luminaires, or wiring enclosures within luminaires used in lieu of boxes in accordance with 300.15(B), shall be supported by rigid or intermediate metal conduit stems. For stems longer than 450 mm (18 in.), the stems shall be connected to the wiring system with flexible fittings suitable for the location. At the luminaire end, the conduit(s) shall be threaded wrenchtight into the box or wiring enclosure, or into hubs identified for the purpose.

 Where supported by only a single conduit, the threaded joints shall be prevented from loosening by the use of setscrews or other effective means, or the luminaire, at any point, shall be at least 2.5 m (8 ft) above grade or standing area and at least 900 mm (3 ft) measured horizontally to the 2.5 m (8 ft) elevation from windows, doors, porches, fire escapes, or similar locations. A luminaire supported by a single conduit shall not exceed 300 mm (12 in.) in any horizontal direction from the point of conduit entry.

314.24 Depth of Boxes. Outlet and device boxes shall have sufficient depth to allow equipment installed within them to be mounted properly and without likelihood of damage to conductors within the box.

(A) Outlet Boxes Without Enclosed Devices or Utilization Equipment. Outlet boxes that do not enclose devices or utilization equipment shall have a minimum internal depth of 12.7 mm (1/2 in.).

(B) Outlet and Device Boxes with Enclosed Devices or Utilization Equipment. Outlet and device boxes that enclose devices or utilization equipment shall have a minimum internal depth that accommodates the rearward projection of the equipment and the size of the conductors that supply the equipment. The internal depth shall include, where used, that of any extension boxes, plaster rings, or raised covers. The internal depth shall comply with all applicable provisions of (B)(1) through (B)(5).

(1) Large Equipment. Boxes that enclose devices or utilization equipment that projects more than 48 mm (1-7/8 in.) rearward from the mounting plane of the box shall have a depth that is not less than the depth of the equipment plus 6 mm (1/4 in.).

(2) Conductors Larger Than 4 AWG. Boxes that enclose devices or utilization equipment supplied by conductors larger than 4 AWG shall be identified for their specific function.

Exception to (2): Devices or utilization equipment supplied by conductors larger than 4 AWG shall be permitted to be mounted on or in junction and pull boxes larger than 1650 cm³ (100 in.³) if the spacing at the terminals meets the requirements of 312.6.

(3) Conductors 8, 6, or 4 AWG. Boxes that enclose devices or utilization equipment supplied by 8, 6, or 4 AWG conductors shall have an internal depth that is not less than 52.4 mm (2-1/16 in.).

(4) Conductors 12 or 10 AWG. Boxes that enclose devices or utilization equipment supplied by 12 or 10 AWG conductors shall have an internal depth that is not less than 30.2 mm (1-3/16 in.). Where the equipment projects rearward from the mounting plane of the box by more than 25 mm (1 in.), the box shall have a depth not less than that of the equipment plus 6 mm (1/4 in.).

(5) Conductors 14 AWG and Smaller. Boxes that enclose devices or utilization equipment supplied by 14 AWG or smaller conductors shall have a depth that is not less than 23.8 mm (15/16 in.).

Exception to (1) through (5): Devices or utilization equipment that is listed to be installed with specified boxes shall be permitted.

314.25 Covers and Canopies. In completed installations, each box shall have a cover, faceplate, lampholder, or luminaire canopy, except where the installation complies with 410.24(B).

(A) Nonmetallic or Metal Covers and Plates. Nonmetallic or metal covers and plates shall be permitted. Where metal covers or plates are used, they shall comply with the grounding requirements of 250.110.

(B) Exposed Combustible Wall or Ceiling Finish. Where a luminaire canopy or pan is used, any combustible wall or ceiling finish exposed between the edge of the canopy or pan and the outlet box shall be covered with noncombustible material.

(C) Flexible Cord Pendants. Covers of outlet boxes and conduit bodies having holes through which flexible cord pendants pass shall be provided with bushings designed for the purpose or shall have smooth, well-rounded surfaces on which the cords may bear. So-called hard rubber or composition bushings shall not be used.

314.27 Outlet Boxes.

(A) Boxes at Luminaire or Lampholder Outlets. Outlet boxes or fittings designed for the support of luminaires and lampholders, and installed as required by 314.23, shall be permitted to support a luminaire or lampholder.

(1) Wall Outlets. Boxes used at luminaire or lampholder outlets in a wall shall be marked on the interior of the box to indicate the maximum weight of the luminaire that is permitted to be supported by the box in the wall, if other than 23 kg (50 lb).

Exception: A wall-mounted luminaire or lampholder weighing not more than 3 kg (6 lb) shall be permitted to be supported on other boxes or plaster rings that are secured to other boxes, provided the luminaire or its supporting yoke, or

*the lampholder, is secured to the box with no fewer than two
No. 6 or larger screws.*

(2) Ceiling Outlets. At every outlet used exclusively for
lighting, the box shall be designed or installed so that a luminaire or lampholder may be attached. Boxes shall be required to support a luminaire weighing a minimum of 23 kg
(50 lb). A luminaire that weighs more than 23 kg (50 lb) shall
be supported independently of the outlet box, unless the outlet box is listed and marked for the maximum weight to be
supported.

(B) Floor Boxes. Boxes listed specifically for this application shall be used for receptacles located in the floor.

(C) Boxes at Ceiling-Suspended (Paddle) Fan Outlets.
Outlet boxes or outlet box systems used as the sole support
of a ceiling-suspended (paddle) fan shall be listed, shall be
marked by their manufacturer as suitable for this purpose,
and shall not support ceiling-suspended (paddle) fans that
weigh more than 32 kg (70 lb). For outlet boxes or outlet box
systems designed to support ceiling-suspended (paddle)
fans that weigh more than 16 kg (35 lb), the required marking
shall include the maximum weight to be supported.

Where spare, separately switched, ungrounded conductors are provided to a ceiling mounted outlet box, in a
location acceptable for a ceiling-suspended (paddle) fan in
single or multi-family dwellings, the outlet box or outlet box
system shall be listed for sole support of a ceiling-suspended
(paddle) fan.

314.28 Pull and Junction Boxes and Conduit Bodies.
Boxes and conduit bodies used as pull or junction boxes shall
comply with 314.28(A) through (E).

(A) Minimum Size. For raceways containing conductors
of 4 AWG or larger that are required to be insulated, and
for cables containing conductors of 4 AWG or larger, the
minimum dimensions of pull or junction boxes installed in a
raceway or cable run shall comply with (A)(1) through (A)(3).
Where an enclosure dimension is to be calculated based on
the diameter of entering raceways, the diameter shall be the

metric designator (trade size) expressed in the units of measurement employed.

(1) Straight Pulls. In straight pulls, the length of the box or conduit body shall not be less than eight times the metric designator (trade size) of the largest raceway.

(2) Angle or U Pulls, or Splices. Where splices or where angle or U pulls are made, the distance between each raceway entry inside the box or conduit body and the opposite wall of the box or conduit body shall not be less than six times the metric designator (trade size) of the largest raceway in a row. This distance shall be increased for additional entries by the amount of the sum of the diameters of all other raceway entries in the same row on the same wall of the box. Each row shall be calculated individually, and the single row that provides the maximum distance shall be used.

Exception: Where a raceway or cable entry is in the wall of a box or conduit body opposite a removable cover, the distance from that wall to the cover shall be permitted to comply with the distance required for one wire per terminal in Table 312.6(A).

The distance between raceway entries enclosing the same conductor shall not be less than six times the metric designator (trade size) of the larger raceway.

When transposing cable size into raceway size in 314.28(A)(1) and (A)(2), the minimum metric designator (trade size) raceway required for the number and size of conductors in the cable shall be used.

(3) Smaller Dimensions. Boxes or conduit bodies of dimensions less than those required in 314.28(A)(1) and (A)(2) shall be permitted for installations of combinations of conductors that are less than the maximum conduit or tubing fill (of conduits or tubing being used) permitted by Table 1 of Chapter 9, provided the box or conduit body has been listed for, and is permanently marked with, the maximum number and maximum size of conductors permitted.

314.29 Boxes, Conduit Bodies, and Handhole Enclosures to Be Accessible. Boxes, conduit bodies, and handhole enclosures shall be installed so that the wiring contained in them can be rendered accessible without removing any part of the building or, in underground circuits, without excavating sidewalks, paving, earth, or other substance that is to be used to establish the finished grade.

Exception: Listed boxes and handhole enclosures shall be permitted where covered by gravel, light aggregate, or non-cohesive granulated soil if their location is effectively identified and accessible for excavation.

314.30 Handhole Enclosures. Handhole enclosures shall be designed and installed to withstand all loads likely to be imposed on them. They shall be identified for use in underground systems.

(A) Size. Handhole enclosures shall be sized in accordance with 314.28(A) for conductors operating at 600 volts or below, and in accordance with 314.71 for conductors operating at over 600 volts. For handhole enclosures without bottoms where the provisions of 314.28(A)(2), Exception, or 314.71(B)(1), Exception No. 1, apply, the measurement to the removable cover shall be taken from the end of the conduit or cable assembly.

(B) Wiring Entries. Underground raceways and cable assemblies entering a handhole enclosure shall extend into the enclosure, but they shall not be required to be mechanically connected to the enclosure.

(C) Enclosed Wiring. All enclosed conductors and any splices or terminations, if present, shall be listed as suitable for wet locations.

(D) Covers. Handhole enclosure covers shall have an identifying mark or logo that prominently identifies the function of the enclosure, such as "electric." Handhole enclosure covers shall require the use of tools to open, or they shall weigh over 45 kg (100 lb). Metal covers and other exposed conductive surfaces shall be bonded in accordance with 250.92 if the conductors in the handhole are service conductors, or in accordance with 250.96(A) if the conductors in the handhole are feeder or branch-circuit conductors.

CHAPTER 13

CABLES

INTRODUCTION

This chapter contains provisions from Article 334—Nonmetallic-Sheathed Cable (Types NM, NMC, and NMS), Article 330—Metal Clad Cable (Type MC), Article 338—Service Entrance Cable (Types SE and USE), and Article 340—Underground Feeder and Branch-Circuit Cable (Type UF). Contained in each article are provisions pertaining to the use and installation for the cable listed. Covered in Part I of each article are the cable's scope and definition. Other topics covered in most of these articles include: uses (permitted and not permitted), bending radius, and ampacity.

While service-entrance cable used as service-entrance conductors must be installed as required by Article 230 where used for interior wiring it must comply with many of the installation requirements in Article 334. Underground feeder and branch-circuit cable installed as nonmetallic-sheathed cable must also comply with the provisions of Article 334.

Although the specifications within each article may be different, the arrangement and numbering in each article follow the same format. For example, the uses permitted for each cable are located in 3__.10, and the uses not permitted are located in 3__.12.

ARTICLE 334

Nonmetallic-Sheathed Cable: Types NM, NMC, and NMS

I. GENERAL

334.1 Scope. This article covers the use, installation, and construction specifications of nonmetallic-sheathed cable.

334.2 Definitions.

Nonmetallic-Sheathed Cable. A factory assembly of two or more insulated conductors enclosed within an overall nonmetallic jacket.

Type NM. Insulated conductors enclosed within an overall nonmetallic jacket.

Type NMC. Insulated conductors enclosed within an overall, corrosion resistant, nonmetallic jacket.

Type NMS. Insulated power or control conductors with signaling, data, and communications conductors within an overall nonmetallic jacket.

334.6 Listed. Type NM, Type NMC, and Type NMS cables shall be listed.

II. INSTALLATION

334.10 Uses Permitted. Type NM, Type NMC, and Type NMS cables shall be permitted to be used in the following:

(1) One- and two-family dwellings and their attached or detached garages, and their storage buildings.
(2) Multifamily dwellings permitted to be of Types III, IV, and V construction except as prohibited in 334.12.

(A) Type NM. Type NM cable shall be permitted as follows:

(1) For both exposed and concealed work in normally dry locations except as prohibited in 334.10(3)

(2) To be installed or fished in air voids in masonry block or tile walls

(B) Type NMC. Type NMC cable shall be permitted as follows:

(1) For both exposed and concealed work in dry, moist, damp, or corrosive locations, except as prohibited by 334.10(3)
(2) In outside and inside walls of masonry block or tile
(3) In a shallow chase in masonry, concrete, or adobe protected against nails or screws by a steel plate at least 1.59 mm (1/16 in.) thick and covered with plaster, adobe, or similar finish

(C) Type NMS. Type NMS cable shall be permitted as follows:

(1) For both exposed and concealed work in normally dry locations except as prohibited by 334.10(3)
(2) To be installed or fished in air voids in masonry block or tile walls

334.12 Uses Not Permitted.

(A) Types NM, NMC, and NMS. Types NM, NMC, and NMS cables shall not be permitted as follows:

(1) In any dwelling or structure not specifically permitted in 334.10(1), (2), and (3)
(2) Exposed in dropped or suspended ceilings in other than one- and two-family and multifamily dwellings
(3) As service-entrance cable
(9) Embedded in poured cement, concrete, or aggregate

(B) Types NM and NMS. Types NM and NMS cables shall not be used under the following conditions or in the following locations:

(1) Where exposed to corrosive fumes or vapors
(2) Where embedded in masonry, concrete, adobe, fill, or plaster
(3) In a shallow chase in masonry, concrete, or adobe and covered with plaster, adobe, or similar finish
(4) In wet or damp locations

334.15 Exposed Work. In exposed work, except as provided in 300.11(A), cable shall be installed as specified in 334.15(A) through (C).

(A) To Follow Surface. Cable shall closely follow the surface of the building finish or of running boards.

(B) Protection from Physical Damage. Cable shall be protected from physical damage where necessary by rigid metal conduit, intermediate metal conduit, electrical metallic tubing, Schedule 80 PVC conduit, Type RTRC marked with the suffix -XW, or other approved means. Where passing through a floor, the cable shall be enclosed in rigid metal conduit, intermediate metal conduit, electrical metallic tubing, Schedule 80 PVC conduit, Type RTRC marked with the suffix -XW, or other approved means extending at least 150 mm (6 in.) above the floor.

Type NMC cable installed in shallow chases or grooves in masonry, concrete, or adobe shall be protected in accordance with the requirements in 300.4(F) and covered with plaster, adobe, or similar finish.

(C) In Unfinished Basements and Crawl Spaces. Where cable is run at angles with joists in unfinished basements and crawl spaces, it shall be permissible to secure cables not smaller than two 6 AWG or three 8 AWG conductors directly to the lower edges of the joists. Smaller cables shall be run either through bored holes in joists or on running boards. Nonmetallic-sheathed cable installed on the wall of an unfinished basement shall be permitted to be installed in a listed conduit or tubing or shall be protected in accordance with 300.4. Conduit or tubing shall be provided with a suitable insulating bushing or adapter at the point the cable enters the raceway. The sheath of the nonmetallic-sheathed cable shall extend through the conduit or tubing and into the outlet or device box not less than 6 mm (1/4 in.). The cable shall be secured within 300 mm (12 in.) of the point where the cable enters the conduit or tubing. Metal conduit, tubing, and metal outlet boxes shall be connected to an equipment grounding conductor complying with the provisions of 250.86 and 250.148.

334.17 Through or Parallel to Framing Members. Types NM, NMC, or NMS cable shall be protected in accordance with 300.4 where installed through or parallel to framing members. Grommets used as required in 300.4(B)(1) shall remain in place and be listed for the purpose of cable protection.

334.23 In Accessible Attics. The installation of cable in accessible attics or roof spaces shall also comply with 320.23.

[**320.23 In Accessible Attics.** Type AC cables in accessible attics or roof spaces shall be installed as specified in 320.23(A) and (B).

(A) Cables Run Across the Top of Floor Joists. Where run across the top of floor joists, or within 2.1 m (7 ft) of the floor or floor joists across the face of rafters or studding, the cable shall be protected by substantial guard strips that are at least as high as the cable. Where this space is not accessible by permanent stairs or ladders, protection shall only be required within 1.8 m (6 ft) of the nearest edge of the scuttle hole or attic entrance.

(B) Cable Installed Parallel to Framing Members. Where the cable is installed parallel to the sides of rafters, studs, or ceiling or floor joists, neither guard strips nor running boards shall be required, and the installation shall also comply with 300.4(D).]

334.24 Bending Radius. Bends in Types NM, NMC, and NMS cable shall be so made that the cable will not be damaged. The radius of the curve of the inner edge of any bend during or after installation shall not be less than five times the diameter of the cable.

334.30 Securing and Supporting. Nonmetallic-sheathed cable shall be supported and secured by staples, cable ties, straps, hangers, or similar fittings designed and installed so as not to damage the cable, at intervals not exceeding 1.4 m (4-1/2 ft) and within 300 mm (12 in.) of every outlet box, junction box, cabinet, or fitting. Flat cables shall not be stapled on edge.

Sections of cable protected from physical damage by raceway shall not be required to be secured within the raceway.

(A) Horizontal Runs Through Holes and Notches. In other than vertical runs, cables installed in accordance with 300.4 shall be considered to be supported and secured where such support does not exceed 1.4-m (4-1/2-ft) intervals and the nonmetallic-sheathed cable is securely fastened in place by an approved means within 300 mm (12 in.) of each box, cabinet, conduit body, or other nonmetallic-sheathed cable termination.

(B) Unsupported Cables. Nonmetallic-sheathed cable shall be permitted to be unsupported where the cable:

(1) Is fished between access points through concealed spaces in finished buildings or structures and supporting is impracticable.

(2) Is not more than 1.4 m (4-1/2 ft) from the last point of cable support to the point of connection to a luminaire or other piece of electrical equipment and the cable and point of connection are within an accessible ceiling.

(C) Wiring Device Without a Separate Outlet Box. A wiring device identified for the use, without a separate outlet box, and incorporating an integral cable clamp shall be permitted where the cable is secured in place at intervals not exceeding 1.4 m (4-1/2 ft) and within 300 mm (12 in.) from the wiring device wall opening, and there shall be at least a 300 mm (12 in.) loop of unbroken cable or 150 mm (6 in.) of a cable end available on the interior side of the finished wall to permit replacement.

334.40 Boxes and Fittings.

(A) Boxes of Insulating Material. Nonmetallic outlet boxes shall be permitted as provided by 314.3.

(B) Devices of Insulating Material. Switch, outlet, and tap devices of insulating material shall be permitted to be used without boxes in exposed cable wiring and for rewiring in existing buildings where the cable is concealed and fished. Openings in such devices shall form a close fit around the outer covering of the cable, and the device shall fully enclose the part of the cable from which any part of the covering has been removed. Where connections to conductors are by binding-screw terminals, there shall be available as many terminals as conductors.

(C) Devices with Integral Enclosures. Wiring devices with integral enclosures identified for such use shall be permitted as provided by 300.15(E).

334.80 Ampacity. The ampacity of Types NM, NMC, and NMS cable shall be determined in accordance with 310.15. The allowable ampacity shall not exceed that of a 60°C (140°F) rated conductor. The 90°C (194°F) rating shall be permitted to be used for ampacity adjustment and correction calculations, provided the final derated ampacity does not exceed that of a 60°C (140°F) rated conductor. The ampacity of Types NM, NMC, and NMS cable installed in cable tray shall be determined in accordance with 392.80(A).

Where more than two NM cables containing two or more current-carrying conductors are installed, without maintaining spacing between the cables, through the same opening in wood framing that is to be sealed with thermal insulation, caulk, or sealing foam, the allowable ampacity of each conductor shall be adjusted in accordance with Table 310.15(B)(3)(a) and the provisions of 310.15(A)(2), Exception, shall not apply.

Where more than two NM cables containing two or more current-carrying conductors are installed in contact with thermal insulation without maintaining spacing between cables, the allowable ampacity of each conductor shall be adjusted in accordance with Table 310.15(B)(3)(a).

ARTICLE 330
Metal-Clad Cable: Type MC

I. GENERAL

330.1 Scope. This article covers the use, installation, and construction specifications of metal-clad cable, Type MC.

330.2 Definition.

Metal Clad Cable, Type MC. A factory assembly of one or more insulated circuit conductors with or without optical fiber members enclosed in an armor of interlocking metal tape, or a smooth or corrugated metallic sheath.

II. INSTALLATION

330.10 Uses Permitted.

(A) General Uses. Type MC cable shall be permitted as follows:

(1)   For services, feeders, and branch circuits.
(2)   For power, lighting, control, and signal circuits.
(3)   Indoors or outdoors.
(4)   Exposed or concealed.
(5)   To be direct buried where identified for such use.
(7)   In any raceway.
(8)   As aerial cable on a messenger.
(10) In dry locations and embedded in plaster finish on brick or other masonry except in damp or wet locations.
(11) In wet locations where any of the following conditions are met:

   a. The metallic covering is impervious to moisture.
   b. A moisture-impervious jacket is provided under the metal covering.
   c. The insulated conductors under the metallic covering are listed for use in wet locations, and a corrosion-resistant jacket is provided over the metallic sheath.

(12) Where single-conductor cables are used, all phase conductors and, where used, the grounded conductor shall be grouped together to minimize induced voltage on the sheath.

(B) Specific Uses. Type MC cable shall be permitted to be installed in compliance with Parts II and III of Article 725 and 770.133 as applicable and in accordance with 330.10(B)(1) through (B)(4).

(2) Direct Buried. Direct-buried cable shall comply with 300.5 or 300.50, as appropriate.

(3) Installed as Service-Entrance Cable. Type MC cable installed as service-entrance cable shall be permitted in accordance with 230.43.

(4) Installed Outside of Buildings or Structures or as Aerial Cable. Type MC cable installed outside of buildings or structures or as aerial cable shall comply with 225.10, 396.10, and 396.12.

> Informational Note: The "Uses Permitted" is not an all-inclusive list.

330.12 Uses Not Permitted. Type MC cable shall not be used under either of the following conditions:

(1) Where subject to physical damage
(2) Where exposed to any of the destructive corrosive conditions in (a) or (b), unless the metallic sheath or armor is resistant to the conditions or is protected by material resistant to the conditions:

    a. Direct buried in the earth or embedded in concrete unless identified for direct burial

    b. Exposed to cinder fills, strong chlorides, caustic alkalis, or vapors of chlorine or of hydrochloric acids

330.17 Through or Parallel to Framing Members. Type MC cable shall be protected in accordance with 300.4(A), (C), and (D) where installed through or parallel to framing members.

330.23 In Accessible Attics. The installation of Type MC cable in accessible attics or roof spaces shall also comply with 320.23. (See 334.23)

330.24 Bending Radius. Bends in Type MC cable shall be so made that the cable will not be damaged. The radius of the curve of the inner edge of any bend shall not be less than required in 330.24(A) through (C).

(A) Smooth Sheath.

(1) Ten times the external diameter of the metallic sheath for cable not more than 19 mm (3/4 in.) in external diameter

(2) Twelve times the external diameter of the metallic sheath for cable more than 19 mm (3/4 in.) but not more than 38 mm (1-1/2 in.) in external diameter

(3) Fifteen times the external diameter of the metallic sheath for cable more than 38 mm (1-1/2 in.) in external diameter

(B) Interlocked-Type Armor or Corrugated Sheath. Seven times the external diameter of the metallic sheath.

(C) Shielded Conductors. Twelve times the overall diameter of one of the individual conductors or seven times the overall diameter of the multiconductor cable, whichever is greater.

330.30 Securing and Supporting.

(A) General. Type MC cable shall be supported and secured by staples, cable ties, straps, hangers, or similar fittings or other approved means designed and installed so as not to damage the cable.

(B) Securing. Unless otherwise provided, cables shall be secured at intervals not exceeding 1.8 m (6 ft). Cables containing four or fewer conductors sized no larger than 10 AWG shall be secured within 300 mm (12 in.) of every box, cabinet, fitting, or other cable termination.

(C) Supporting. Unless otherwise provided, cables shall be supported at intervals not exceeding 1.8 m (6 ft).

Horizontal runs of Type MC cable installed in wooden or metal framing members or similar supporting means shall be considered supported and secured where such support does not exceed 1.8-m (6-ft) intervals.

(D) Unsupported Cables. Type MC cable shall be permitted to be unsupported where the cable:

(1) Is fished between access points through concealed spaces in finished buildings or structures and supporting is impractical; or

(2) Is not more than 1.8 m (6 ft) in length from the last point of cable support to the point of connection to luminaires or other electrical equipment and the cable and point

of connection are within an accessible ceiling. For the purpose of this section, Type MC cable fittings shall be permitted as a means of cable support.

330.40 Boxes and Fittings. Fittings used for connecting Type MC cable to boxes, cabinets, or other equipment shall be listed and identified for such use.

330.80 Ampacity. The ampacity of Type MC cable shall be determined in accordance with 310.15 or 310.60 for 14 AWG and larger conductors and in accordance with Table 402.5 for 18 AWG and 16 AWG conductors. The installation shall not exceed the temperature ratings of terminations and equipment.

ARTICLE 338
Service-Entrance Cable: Types SE and USE

I. GENERAL

338.1 Scope. This article covers the use, installation, and construction specifications of service-entrance cable.

338.2 Definitions.

Service-Entrance Cable. A single conductor or multiconductor assembly provided with or without an overall covering, primarily used for services, and of the following types:

Type SE. Service-entrance cable having a flame-retardant, moisture-resistant covering.

Type USE. Service-entrance cable, identified for underground use, having a moisture-resistant covering, but not required to have a flame-retardant covering.

II. INSTALLATION

338.10 Uses Permitted.

(A) Service-Entrance Conductors. Service-entrance cable shall be permitted to be used as service-entrance conductors and shall be installed in accordance with 230.6, 230.7, and Parts II, III, and IV of Article 230.

(B) Branch Circuits or Feeders.

(1) Grounded Conductor Insulated. Type SE service-entrance cables shall be permitted in wiring systems where all of the circuit conductors of the cable are of the thermoset or thermoplastic type.

(2) Use of Uninsulated Conductor. Type SE service-entrance cable shall be permitted for use where the insulated conductors are used for circuit wiring and the uninsulated conductor is used only for equipment grounding purposes.

Exception: In existing installations, uninsulated conductors shall be permitted as a grounded conductor in accordance with 250.32 and 250.140, where the uninsulated grounded conductor of the cable originates in service equipment, and with 225.30 through 225.40.

(3) Temperature Limitations. Type SE service-entrance cable used to supply appliances shall not be subject to conductor temperatures in excess of the temperature specified for the type of insulation involved.

(4) Installation Methods for Branch Circuits and Feeders.
    (a) *Interior Installations.* In addition to the provisions of this article, Type SE service-entrance cable used for interior wiring shall comply with the installation requirements of Part II of Article 334, excluding 334.80. Where installed in thermal insulation, the ampacity shall be in accordance with the 60°C (140°F) conductor temperature rating. The maximum conductor temperature rating shall be permitted to be used for ampacity adjustment and correction purposes, if the final derated ampacity does not exceed that for a 60°C (140°F) rated conductor.
    (b) *Exterior Installations.* In addition to the provisions of this article, service-entrance cable used for feeders or branch circuits, where installed as exterior wiring, shall be installed in accordance with Part I of Article 225. The cable shall be supported in accordance with 334.30. Type USE cable installed as underground feeder and branch circuit cable shall comply with Part II of Article 340.

338.12 Uses Not Permitted.

(A) Service-Entrance Cable. Service-entrance cable (SE) shall not be used under the following conditions or in the following locations:

(1) Where subject to physical damage unless protected in accordance with 230.50(B)
(2) Underground with or without a raceway
(3) For exterior branch circuits and feeder wiring unless the installation complies with the provisions of Part I of Article 225 and is supported in accordance with 334.30 or is used as messenger-supported wiring as permitted in Part II of Article 396

(B) Underground Service-Entrance Cable. Underground service-entrance cable (USE) shall not be used under the following conditions or in the following locations:

(1) For interior wiring
(2) For aboveground installations except where USE cable emerges from the ground and is terminated in an enclosure at an outdoor location and the cable is protected in accordance with 300.5(D)
(3) As aerial cable unless it is a multiconductor cable identified for use aboveground and installed as messenger-supported wiring in accordance with 225.10 and Part II of Article 396

338.24 Bending Radius. Bends in Types USE and SE cable shall be so made that the cable will not be damaged. The radius of the curve of the inner edge of any bend, during or after installation, shall not be less than five times the diameter of the cable.

ARTICLE 340
Underground Feeder and Branch-Circuit Cable: Type UF

I. GENERAL

340.1 Scope. This article covers the use, installation, and construction specifications for underground feeder and branch-circuit cable, Type UF.

340.2 Definition.

Underground Feeder and Branch-Circuit Cable, Type UF. A factory assembly of one or more insulated conductors with an integral or an overall covering of nonmetallic material suitable for direct burial in the earth.

340.6 Listing Requirements. Type UF cable shall be listed.

II. INSTALLATION

340.10 Uses Permitted. Type UF cable shall be permitted as follows:

(1) For use underground, including direct burial in the earth. For underground requirements, see 300.5.

(2) As single-conductor cables. Where installed as single-conductor cables, all conductors of the feeder grounded conductor or branch circuit, including the grounded conductor and equipment grounding conductor, if any, shall be installed in accordance with 300.3.

(3) For wiring in wet, dry, or corrosive locations under the recognized wiring methods of this *Code*.

(4) Installed as nonmetallic-sheathed cable. Where so installed, the installation and conductor requirements shall comply with Parts II and III of Article 334 and shall be of the multiconductor type.

(5) For solar photovoltaic systems in accordance with 690.31.

(6) As single-conductor cables as the nonheating leads for heating cables as provided in 424.43.

340.12 Uses Not Permitted. Type UF cable shall not be used as follows:

(1) As service-entrance cable

(8) Embedded in poured cement, concrete, or aggregate, except where embedded in plaster as nonheating leads where permitted in 424.43.

(9) Where exposed to direct rays of the sun, unless identified as sunlight resistant

(10) Where subject to physical damage

(11) As overhead cable, except where installed as messenger-supported wiring in accordance with Part II of Article 396

340.24 Bending Radius. Bends in Type UF cable shall be so made that the cable is not damaged. The radius of the curve of the inner edge of any bend shall not be less than five times the diameter of the cable.

340.80 Ampacity. The ampacity of Type UF cable shall be that of 60°C (140°F) conductors in accordance with 310.15.

CHAPTER 14

RACEWAYS

INTRODUCTION

This chapter contains provisions from Article 358—Electrical Metallic Tubing (Type EMT), Article 344—Rigid Metal Conduit (Type RMC), Article 352—Rigid Polyvinyl Chloride Conduit (Type PVC), Article 362—Electrical Nonmetallic Tubing (Type ENT), Article 348—Flexible Metal Conduit (Type FMC), Article 350—Liquidtight Flexible Metal Conduit (Type LFMC), and Article 356—Liquidtight Flexible Nonmetallic Conduit (Type LFNC). Contained in each article are provisions pertaining to the use and installation for the conduit or tubing listed. Part I of each article covers the raceway's scope and definition. Other topics covered in most of these articles include: uses (permitted and not permitted), size (minimum and maximum), number of conductors, bends (how made and number in one run), securing and supporting, and grounding.

Although the requirements in each article may be different, the arrangement and numbering in each article follow the same format. For example, in each article, 3__.10 covers uses permitted, 3__.12 covers uses not permitted, 3__.20 covers size, 3__.22 covers number of conductors, 3__.30 covers securing and supporting, and 3__.60 covers grounding.

ARTICLE 358
Electrical Metallic Tubing: Type EMT

I. GENERAL

358.1 Scope. This article covers the use, installation, and construction specifications for electrical metallic tubing (EMT) and associated fittings.

358.2 Definition.

Electrical Metallic Tubing (EMT). An unthreaded thinwall raceway of circular cross section designed for the physical protection and routing of conductors and cables and for use as an equipment grounding conductor when installed utilizing appropriate fittings. EMT is generally made of steel (ferrous) with protective coatings or aluminum (nonferrous).

358.6 Listing Requirements. EMT, factory elbows, and associated fittings shall be listed.

II. INSTALLATION

358.10 Uses Permitted.

(A) Exposed and Concealed. The use of EMT shall be permitted for both exposed and concealed work.

(B) Corrosion Protection. Ferrous or nonferrous EMT, elbows, couplings, and fittings shall be permitted to be installed in concrete, in direct contact with the earth, or in areas subject to severe corrosive influences where protected by corrosion protection and approved as suitable for the condition.

(C) Wet Locations. All supports, bolts, straps, screws, and so forth shall be of corrosion-resistant materials or protected against corrosion by corrosion-resistant materials.

358.12 Uses Not Permitted. EMT shall not be used under the following conditions:

(1) Where, during installation or afterward, it will be subject to severe physical damage.
(2) Where protected from corrosion solely by enamel.

(3) In cinder concrete or cinder fill where subject to permanent moisture unless protected on all sides by a layer of noncinder concrete at least 50 mm (2 in.) thick or unless the tubing is at least 450 mm (18 in.) under the fill.

(5) For the support of luminaires or other equipment except conduit bodies no larger than the largest trade size of the tubing.

358.24 Bends—How Made. Bends shall be made so that the tubing is not damaged and the internal diameter of the tubing is not effectively reduced. The radius of the curve of any field bend to the centerline of the tubing shall be not less than shown in Table 2, Chapter 9 for one-shot and full shoe benders.

358.26 Bends—Number in One Run. There shall not be more than the equivalent of four quarter bends (360 degrees total) between pull points, for example, conduit bodies and boxes.

358.28 Reaming and Threading.

(A) Reaming. All cut ends of EMT shall be reamed or otherwise finished to remove rough edges.

358.30 Securing and Supporting. EMT shall be installed as a complete system in accordance with 300.18 and shall be securely fastened in place and supported in accordance with 358.30(A) and (B).

(A) Securely Fastened. EMT shall be securely fastened in place at least every 3 m (10 ft). In addition, each EMT run between termination points shall be securely fastened within 900 mm (3 ft) of each outlet box, junction box, device box, cabinet, conduit body, or other tubing termination.

Exception No. 1: Fastening of unbroken lengths shall be permitted to be increased to a distance of 1.5 m (5 ft) where structural members do not readily permit fastening within 900 mm (3 ft).

Exception No. 2: For concealed work in finished buildings or prefinished wall panels where such securing is impracticable, unbroken lengths (without coupling) of EMT shall be permitted to be fished.

(B) Supports. Horizontal runs of EMT supported by openings through framing members at intervals not greater than 3 m (10 ft) and securely fastened within 900 mm (3 ft) of termination points shall be permitted.

358.42 Couplings and Connectors. Couplings and connectors used with EMT shall be made up tight. Where buried in masonry or concrete, they shall be concretetight type. Where installed in wet locations, they shall comply with 314.15.

358.60 Grounding. EMT shall be permitted as an equipment grounding conductor.

ARTICLE 344
Rigid Metal Conduit: Type RMC

I. GENERAL

344.1 Scope. This article covers the use, installation, and construction specifications for rigid metal conduit (RMC) and associated fittings.

344.2 Definition.

Rigid Metal Conduit (RMC). A threadable raceway of circular cross section designed for the physical protection and routing of conductors and cables and for use as an equipment grounding conductor when installed with its integral or associated coupling and appropriate fittings. RMC is generally made of steel (ferrous) with protective coatings or aluminum (nonferrous). Special use types are red brass and stainless steel.

344.6 Listing Requirements. RMC, factory elbows and couplings, and associated fittings shall be listed.

II. INSTALLATION

344.10 Uses Permitted.

(A) Atmospheric Conditions and Occupancies.

(1) Galvanized Steel and Stainless Steel RMC. Galvanized steel and stainless steel RMC shall be permitted under all atmospheric conditions and occupancies.

(2) Red Brass RMC. Red brass RMC shall be permitted to be installed for direct burial and swimming pool applications.

(3) Aluminum RMC. Aluminum RMC shall be permitted to be installed where judged suitable for the environment. Rigid aluminum conduit encased in concrete or in direct contact with the earth shall be provided with approved supplementary corrosion protection.

(4) Ferrous Raceways and Fittings. Ferrous raceways and fittings protected from corrosion solely by enamel shall be permitted only indoors and in occupancies not subject to severe corrosive influences.

(B) Corrosive Environments.

(1) Galvanized Steel, Stainless Steel, and Red Brass RMC, Elbows, Couplings, and Fittings. Galvanized steel, stainless steel, and red brass RMC elbows, couplings, and fittings shall be permitted to be installed in concrete, in direct contact with the earth, or in areas subject to severe corrosive influences where protected by corrosion protection and judged suitable for the condition.

(2) Supplementary Protection of Aluminum RMC. Aluminum RMC shall be provided with approved supplementary corrosion protection where encased in concrete or in direct contact with the earth.

(C) Cinder Fill. Galvanized steel, stainless steel, and red brass RMC shall be permitted to be installed in or under cinder fill where subject to permanent moisture where protected on all sides by a layer of noncinder concrete not less than 50 mm (2 in.) thick; where the conduit is not less than 450 mm (18 in.) under the fill; or where protected by corrosion protection and judged suitable for the condition.

(D) Wet Locations. All supports, bolts, straps, screws, and so forth, shall be of corrosion-resistant materials or protected against corrosion by corrosion-resistant materials.

344.24 Bends—How Made. Bends of RMC shall be so made that the conduit will not be damaged and so that the internal diameter of the conduit will not be effectively reduced. The radius of the curve of any field bend to the centerline of the conduit shall not be less than indicated in Table 2, Chapter 9.

344.26 Bends—Number in One Run. There shall not be more than the equivalent of four quarter bends (360 degrees total) between pull points, for example, conduit bodies and boxes.

344.28 Reaming and Threading. All cut ends shall be reamed or otherwise finished to remove rough edges. Where conduit is threaded in the field, a standard cutting die with a 1 in 16 taper (3/4 in. taper per foot) shall be used.

344.30 Securing and Supporting. RMC shall be installed as a complete system in accordance with 300.18 and shall be securely fastened in place and supported in accordance with 344.30(A) and (B).

(A) Securely Fastened. RMC shall be securely fastened within 900 mm (3 ft) of each outlet box, junction box, device box, cabinet, conduit body, or other conduit termination. Fastening shall be permitted to be increased to a distance of 1.5 m (5 ft) where structural members do not readily permit fastening within 900 mm (3 ft). Where approved, conduit shall not be required to be securely fastened within 900 mm (3 ft) of the service head for above-the-roof termination of a mast.

(B) Supports. RMC shall be supported in accordance with one of the following:

(1) Conduit shall be supported at intervals not exceeding 3 m (10 ft).
(2) The distance between supports for straight runs of conduit shall be permitted in accordance with Table 344.30(B)(2), provided the conduit is made up with threaded couplings and such supports prevent transmission of stresses to termination where conduit is deflected between supports.
(4) Horizontal runs of RMC supported by openings through framing members at intervals not exceeding 3 m (10 ft) and securely fastened within 900 mm (3 ft) of termination points shall be permitted.

344.42 Couplings and Connectors.

(A) Threadless. Threadless couplings and connectors used with conduit shall be made tight. Where buried in masonry or

Table 344.30(B)(2) Supports for Rigid Metal Conduit

Conduit Size		Maximum Distance Between Rigid Metal Conduit Supports	
Metric Designator	Trade Size	m	ft
16–21	1/2–3/4	3.0	10
27	1	3.7	12
35–41	1-1/4–1-1/2	4.3	14
53–63	2–2-1/2	4.9	16
78 and larger	3 and larger	6.1	20

concrete, they shall be the concretetight type. Where installed in wet locations, they shall comply with 314.15. Threadless couplings and connectors shall not be used on threaded conduit ends unless listed for the purpose.

(B) Running Threads. Running threads shall not be used on conduit for connection at couplings.

344.46 Bushings. Where a conduit enters a box, fitting, or other enclosure, a bushing shall be provided to protect the wires from abrasion unless the box, fitting, or enclosure is designed to provide such protection.

Informational Note: See 300.4(G) for the protection of conductors sizes 4 AWG and larger at bushings.

344.60 Grounding. RMC shall be permitted as an equipment grounding conductor.

ARTICLE 352
Rigid Polyvinyl Chloride Conduit: Type PVC

I. GENERAL

352.1 Scope. This article covers the use, installation, and construction specifications for rigid polyvinyl chloride conduit (PVC) and associated fittings.

352.2 Definition.

Rigid Polyvinyl Chloride Conduit (PVC). A rigid nonmetallic conduit of circular cross section, with integral or associated couplings, connectors, and fittings for the installation of electrical conductors and cables.

352.6 Listing Requirements. PVC conduit, factory elbows, and associated fittings shall be listed.

II. INSTALLATION

352.10 Uses Permitted. The use of PVC conduit shall be permitted in accordance with 352.10(A) through (H).

(A) Concealed. PVC conduit shall be permitted in walls, floors, and ceilings.

(B) Corrosive Influences. PVC conduit shall be permitted in locations subject to severe corrosive influences as covered in 300.6 and where subject to chemicals for which the materials are specifically approved.

(C) Cinders. PVC conduit shall be permitted in cinder fill.

(D) Wet Locations. PVC conduit shall be permitted in portions of dairies, laundries, canneries, or other wet locations, and in locations where walls are frequently washed, the entire conduit system, including boxes and fittings used therewith, shall be installed and equipped so as to prevent water from entering the conduit. All supports, bolts, straps, screws, and so forth, shall be of corrosion-resistant materials or be protected against corrosion by approved corrosion-resistant materials.

(E) Dry and Damp Locations. PVC conduit shall be permitted for use in dry and damp locations not prohibited by 352.12.

(F) Exposed. PVC conduit shall be permitted for exposed work. PVC conduit used exposed in areas of physical damage shall be identified for the use.

(G) Underground Installations. For underground installations, PVC shall be permitted for direct burial and underground encased in concrete. See 300.5 and 300.50.

(H) Support of Conduit Bodies. PVC conduit shall be permitted to support nonmetallic conduit bodies not larger than the largest trade size of an entering raceway. These conduit bodies shall not support luminaires or other equipment and shall not contain devices other than splicing devices as permitted by 110.14(B) and 314.16(C)(2).

(I) Insulation Temperature Limitations. Conductors or cables rated at a temperature higher than the listed temperature rating of PVC conduit shall be permitted to be installed in PVC conduit, provided the conductors or cables are not operated at a temperature higher than the listed temperature rating of the PVC conduit.

352.12 Uses Not Permitted. PVC conduit shall not be used under the conditions specified in 352.12(A) through (E).

(B) Support of Luminaires. For the support of luminaires or other equipment not described in 352.10(H).

(C) Physical Damage. Where subject to physical damage unless identified for such use.

(D) Ambient Temperatures. Where subject to ambient temperatures in excess of 50°C (122°F) unless listed otherwise.

352.24 Bends—How Made. Bends shall be so made that the conduit will not be damaged and the internal diameter of the conduit will not be effectively reduced. Field bends shall be made only with bending equipment identified for the purpose. The radius of the curve to the centerline of such bends shall not be less than shown in Table 2, Chapter 9.

352.26 Bends—Number in One Run. There shall not be more than the equivalent of four quarter bends (360 degrees total) between pull points, for example, conduit bodies and boxes.

352.28 Trimming. All cut ends shall be trimmed inside and outside to remove rough edges.

352.30 Securing and Supporting. PVC conduit shall be installed as a complete system as provided in 300.18 and shall be fastened so that movement from thermal expansion or contraction is permitted. PVC conduit shall be securely fastened and supported in accordance with 352.30(A) and (B).

(A) Securely Fastened. PVC conduit shall be securely fastened within 900 mm (3 ft) of each outlet box, junction box, device box, conduit body, or other conduit termination. Conduit listed for securing at other than 900 mm (3 ft) shall be permitted to be installed in accordance with the listing.

(B) Supports. PVC conduit shall be supported as required in Table 352.30. Conduit listed for support at spacings other than as shown in Table 352.30 shall be permitted to be installed in accordance with the listing. Horizontal runs of PVC conduit supported by openings through framing members at intervals not exceeding those in Table 352.30 and securely fastened within 900 mm (3 ft) of termination points shall be permitted.

352.44 Expansion Fittings. Expansion fittings for PVC conduit shall be provided to compensate for thermal expansion and contraction where the length change, in accordance with Table 352.44, is expected to be 6 mm (1/4 in.) or greater in a straight run between securely mounted items such as boxes, cabinets, elbows, or other conduit terminations.

352.46 Bushings. Where a conduit enters a box, fitting, or other enclosure, a bushing or adapter shall be provided to protect the wire from abrasion unless the box, fitting, or enclosure design provides equivalent protection.

Table 352.30 Support of Rigid Polyvinyl Chloride Conduit (PVC)

Conduit Size		Maximum Spacing Between Supports	
Metric Designator	Trade Size	mm or m	ft
16–27	1/2–1	900 mm	3
35–53	1-1/4–2	1.5 m	5
63–78	2-1/2–3	1.8 m	6
91–129	3-1/2–5	2.1 m	7
155	6	2.5 m	8

Table 352.44 Expansion Characteristics of PVC Rigid Nonmetallic Conduit Coefficient of Thermal Expansion = 6.084× 10^{-5} mm/mm/°C (3.38 × 10^{-5} in./in./°F)

Temperature Change (°C)	Length Change of PVC Conduit (mm/m)	Temperature Change (°F)	Length Change of PVC Conduit (in./100 ft)	Temperature Change (°F)	Length Change of PVC Conduit (in./100 ft)
5	0.30	5	0.20	105	4.26
10	0.61	10	0.41	110	4.46
15	0.91	15	0.61	115	4.66
20	1.22	20	0.81	120	4.87
25	1.52	25	1.01	125	5.07
30	1.83	30	1.22	130	5.27
35	2.13	35	1.42	135	5.48
40	2.43	40	1.62	140	5.68
45	2.74	45	1.83	145	5.88
50	3.04	50	2.03	150	6.08
55	3.35	55	2.23	155	6.29
60	3.65	60	2.43	160	6.49
65	3.95	65	2.64	165	6.69
70	4.26	70	2.84	170	6.90
75	4.56	75	3.04	175	7.10
80	4.87	80	3.24	180	7.30
85	5.17	85	3.45	185	7.50
90	5.48	90	3.65	190	7.71
95	5.78	95	3.85	195	7.91
100	6.08	100	4.06	200	8.11

Informational Note: See 300.4(G) for the protection
of conductors 4 AWG and larger at bushings.

352.48 Joints. All joints between lengths of conduit, and
between conduit and couplings, fittings, and boxes, shall be
made by an approved method.

352.60 Grounding. Where equipment grounding is required, a separate equipment grounding conductor shall be
installed in the conduit.

ARTICLE 362
Electrical Nonmetallic Tubing: Type ENT

I. GENERAL

362.1 Scope. This article covers the use, installation, and
construction specifications for electrical nonmetallic tubing
(ENT) and associated fittings.

362.2 Definition.

Electrical Nonmetallic Tubing (ENT). A nonmetallic, pliable, corrugated raceway of circular cross section with integral or associated couplings, connectors, and fittings for
the installation of electrical conductors. ENT is composed of
a material that is resistant to moisture and chemical atmo-
spheres and is flame retardant.

A pliable raceway is a raceway that can be bent by hand
with a reasonable force but without other assistance.

362.6 Listing Requirements. ENT and associated fittings
shall be listed.

II. INSTALLATION

362.10 Uses Permitted. For the purpose of this article, the
first floor of a building shall be that floor that has 50 percent
or more of the exterior wall surface area level with or above
finished grade. One additional level that is the first level and
not designed for human habitation and used only for vehicle

parking, storage, or similar use shall be permitted. The use of ENT and fittings shall be permitted in the following:

(1) In any building not exceeding three floors above grade as follows:

    a. For exposed work, where not prohibited by 362.12

    b. Concealed within walls, floors, and ceilings

(2) In any building exceeding three floors above grade, ENT shall be concealed within walls, floors, and ceilings where the walls, floors, and ceilings provide a thermal barrier of material that has at least a 15-minute finish rating as identified in listings of fire-rated assemblies. The 15-minute-finish-rated thermal barrier shall be permitted to be used for combustible or noncombustible walls, floors, and ceilings.

Exception to (2): Where a fire sprinkler system(s) is installed in accordance with NFPA 13–2010, Standard for the Installation of Sprinkler Systems, on all floors, ENT shall be permitted to be used within walls, floors, and ceilings, exposed or concealed, in buildings exceeding three floors abovegrade.

(3) In locations subject to severe corrosive influences as covered in 300.6 and where subject to chemicals for which the materials are specifically approved.

(4) In concealed, dry, and damp locations not prohibited by 362.12.

(5) Above suspended ceilings where the suspended ceilings provide a thermal barrier of material that has at least a 15-minute finish rating as identified in listings of fire-rated assemblies, except as permitted in 362.10(1)(a).

Exception to (5): ENT shall be permitted to be used above suspended ceilings in buildings exceeding three floors above grade where the building is protected throughout by a fire sprinkler system installed in accordance with NFPA 13–2010, Standard for the Installation of Sprinkler Systems.

(6) Encased in poured concrete, or embedded in a concrete slab on grade where ENT is placed on sand or approved

screenings, provided fittings identified for this purpose are used for connections.

(7) For wet locations indoors as permitted in this section or in a concrete slab on or belowgrade, with fittings listed for the purpose.

(8) Metric designator 16 through 27 (trade size 1/2 through 1) as listed manufactured prewired assembly.

(9) Conductors or cables rated at a temperature higher than the listed temperature rating of ENT shall be permitted to be installed in ENT, if the conductors or cables are not operated at a temperature higher than the listed temperature rating of the ENT.

362.12 Uses Not Permitted. ENT shall not be used in the following:

(2) For the support of luminaires and other equipment

(3) Where subject to ambient temperatures in excess of 50°C (122°F) unless listed otherwise

(4) For direct earth burial

(6) In exposed locations, except as permitted by 362.10(1), 362.10(5), and 362.10(7)

(8) Where exposed to the direct rays of the sun, unless identified as sunlight resistant

(9) Where subject to physical damage

362.24 Bends—How Made. Bends shall be so made that the tubing will not be damaged and the internal diameter of the tubing will not be effectively reduced. Bends shall be permitted to be made manually without auxiliary equipment, and the radius of the curve to the centerline of such bends shall not be less than shown in Table 2, Chapter 9 using the column "Other Bends."

362.26 Bends—Number in One Run. There shall not be more than the equivalent of four quarter bends (360 degrees total) between pull points, for example, conduit bodies and boxes.

362.28 Trimming. All cut ends shall be trimmed inside and outside to remove rough edges.

362.30 Securing and Supporting. ENT shall be installed as a complete system in accordance with 300.18 and shall be securely fastened in place and supported in accordance with 362.30(A) and (B).

(A) Securely Fastened. ENT shall be securely fastened at intervals not exceeding 900 mm (3 ft). In addition, ENT shall be securely fastened in place within 900 mm (3 ft) of each outlet box, device box, junction box, cabinet, or fitting where it terminates.

Exception No. 1: Lengths not exceeding a distance of 1.8 m (6 ft) from a luminaire terminal connection for tap connections to lighting luminaires shall be permitted without being secured.

Exception No. 2: Lengths not exceeding 1.8 m (6 ft) from the last point where the raceway is securely fastened for connections within an accessible ceiling to luminaire(s) or other equipment.

Exception No. 3: For concealed work in finished buildings or prefinished wall panels where such securing is impracticable, unbroken lengths (without coupling) of ENT shall be permitted to be fished.

(B) Supports. Horizontal runs of ENT supported by openings in framing members at intervals not exceeding 900 mm (3 ft) and securely fastened within 900 mm (3 ft) of termination points shall be permitted.

362.46 Bushings. Where a tubing enters a box, fitting, or other enclosure, a bushing or adapter shall be provided to protect the wire from abrasion unless the box, fitting, or enclosure design provides equivalent protection.

Informational Note: See 300.4(G) for the protection of conductors size 4 AWG or larger.

362.48 Joints. All joints between lengths of tubing and between tubing and couplings, fittings, and boxes shall be by an approved method.

362.60 Grounding. Where equipment grounding is required, a separate equipment grounding conductor shall be installed in the raceway in compliance with Article 250, Part VI.

ARTICLE 348
Flexible Metal Conduit: Type FMC

I. GENERAL

348.1 Scope. This article covers the use, installation, and construction specifications for flexible metal conduit (FMC) and associated fittings.

348.2 Definition.

Flexible Metal Conduit (FMC). A raceway of circular cross section made of helically wound, formed, interlocked metal strip.

348.6 Listing Requirements. FMC and associated fittings shall be listed.

II. INSTALLATION

348.10 Uses Permitted. FMC shall be permitted to be used in exposed and concealed locations.

348.12 Uses Not Permitted. FMC shall not be used in the following:

(1) In wet locations
(5) Where exposed to materials having a deteriorating effect on the installed conductors, such as oil or gasoline
(6) Underground or embedded in poured concrete or aggregate
(7) Where subject to physical damage

348.22 Number of Conductors. The number of conductors shall not exceed that permitted by the percentage fill specified in Table 1, Chapter 9, or as permitted in Table 348.22, or for metric designator 12 (trade size 3/8).

Cables shall be permitted to be installed where such use is not prohibited by the respective cable articles. The number of cables shall not exceed the allowable percentage fill specified in Table 1, Chapter 9.

Table 348.22 Maximum Number of Insulated Conductors in Metric Designator 12 (Trade Size 3/8) Flexible Metal Conduit*

Size (AWG)	Types RFH-2, SF-2		Types TF, XHHW, TW		Types TFN, THHN, THWN		Types FEP, FEBP, PF, PGF	
	Fittings Inside Conduit	Fittings Outside Conduit	Fittings Inside Conduit	Fittings Outside Conduit	Fittings Inside Conduit	Fittings Outside Conduit	Fittings Inside Conduit	Fittings Outside Conduit
18	2	3	3	5	5	8	5	8
16	1	2	3	4	4	6	4	6
14	1	2	2	3	3	4	3	4
12	—	—	1	2	2	3	2	3
10	—	—	1	1	1	1	1	2

*In addition, one insulated, covered, or bare equipment grounding conductor of the same size shall be permitted.

348.26 Bends—Number in One Run. There shall not be more than the equivalent of four quarter bends (360 degrees total) between pull points, for example, conduit bodies and boxes.

348.28 Trimming. All cut ends shall be trimmed or otherwise finished to remove rough edges, except where fittings that thread into the convolutions are used.

348.30 Securing and Supporting. FMC shall be securely fastened in place and supported in accordance with 348.30(A) and (B).

(A) Securely Fastened. FMC shall be securely fastened in place by an approved means within 300 mm (12 in.) of each box, cabinet, conduit body, or other conduit termination and shall be supported and secured at intervals not to exceed 1.4 m (4-1/2 ft).

Exception No. 1: Where FMC is fished between access points through concealed spaces in finished buildings or structures and supporting is impracticable.

Exception No. 2: Where flexibility is necessary after installation, lengths from the last point where the raceway is securely fastened shall not exceed the following:

(1) 900 mm (3 ft) for metric designators 16 through 35 (trade sizes 1/2 through 1-1/4)
(2) 1200 mm (4 ft) for metric designators 41 through 53 (trade sizes 1-1/2 through 2)
(3) 1500 mm (5 ft) for metric designator 63 (trade size 2-1/2) and larger

Exception No. 3: Lengths not exceeding 1.8 m (6 ft) from a luminaire terminal connection for tap connections to luminaires as permitted in 410.117(C).

Exception No. 4: Lengths not exceeding 1.8 m (6 ft) from the last point where the raceway is securely fastened for connections within an accessible ceiling to luminaire(s) or other equipment.

(B) Supports. Horizontal runs of FMC supported by openings through framing members at intervals not greater than

1.4 m (4-1/2 ft) and securely fastened within 300 mm (12 in.) of termination points shall be permitted.

348.42 Couplings and Connectors. Angle connectors shall not be concealed.

348.60 Grounding and Bonding. If used to connect equipment where flexibility is necessary to minimize the transmission of vibration from equipment or to provide flexibility for equipment that requires movement after installation, an equipment grounding conductor shall be installed.

Where flexibility is not required after installation, FMC shall be permitted to be used as an equipment grounding conductor when installed in accordance with 250.118(5).

Where required or installed, equipment grounding conductors shall be installed in accordance with 250.134(B).

Where required or installed, equipment bonding jumpers shall be installed in accordance with 250.102.

ARTICLE 350
Liquidtight Flexible Metal Conduit: Type LFMC

I. GENERAL

350.1 Scope. This article covers the use, installation, and construction specifications for liquidtight flexible metal conduit (LFMC) and associated fittings.

350.2 Definition.

Liquidtight Flexible Metal Conduit (LFMC). A raceway of circular cross section having an outer liquidtight, nonmetallic, sunlight-resistant jacket over an inner flexible metal core with associated couplings, connectors, and fittings for the installation of electric conductors.

350.6 Listing Requirements. LFMC and associated fittings shall be listed.

II. INSTALLATION

350.10 Uses Permitted. LFMC shall be permitted to be used in exposed or concealed locations as follows:

(1) Where conditions of installation, operation, or maintenance require flexibility or protection from liquids, vapors, or solids

(3) For direct burial where listed and marked for the purpose

350.12 Uses Not Permitted. LFMC shall not be used as follows:

(1) Where subject to physical damage
(2) Where any combination of ambient and conductor temperature produces an operating temperature in excess of that for which the material is approved

350.26 Bends—Number in One Run. There shall not be more than the equivalent of four quarter bends (360 degrees total) between pull points, for example, conduit bodies and boxes.

350.30 Securing and Supporting. LFMC shall be securely fastened in place and supported in accordance with 350.30(A) and (B).

(A) Securely Fastened. LFMC shall be securely fastened in place by an approved means within 300 mm (12 in.) of each box, cabinet, conduit body, or other conduit termination and shall be supported and secured at intervals not to exceed 1.4 m (4-1/2 ft).

Exception No. 1: Where LFMC is fished between access points through concealed spaces in finished buildings or structures and supporting is impractical.

Exception No. 2: Where flexibility is necessary after installation, lengths from the last point where the raceway is securely fastened shall not exceed the following:

(1) 900 mm (3 ft) for metric designators 16 through 35 (trade sizes 1/2 through 1-1/4)
(2) 1200 mm (4 ft) for metric designators 41 through 53 (trade sizes 1-1/2 through 2)
(3) 1500 mm (5 ft) for metric designator 63 (trade size 2-1/2) and larger

Exception No. 3: Lengths not exceeding 1.8 m (6 ft) from a luminaire terminal connection for tap conductors to luminaires, as permitted in 410.117(C).

Exception No. 4: Lengths not exceeding 1.8 m (6 ft) from the last point where the raceway is securely fastened for connections within an accessible ceiling to luminaire(s) or other equipment.

(B) Supports. Horizontal runs of LFMC supported by openings through framing members at intervals not greater than 1.4 m (4-1/2 ft) and securely fastened within 300 mm (12 in.) of termination points shall be permitted.

350.42 Couplings and Connectors. Angle connectors shall not be concealed.

350.60 Grounding and Bonding. If used to connect equipment where flexibility is necessary to minimize the transmission of vibration from equipment or to provide flexibility for equipment that requires movement after installation, an equipment grounding conductor shall be installed.

Where flexibility is not required after installation, LFMC shall be permitted to be used as an equipment grounding conductor when installed in accordance with 250.118(6).

Where required or installed, equipment grounding conductors shall be installed in accordance with 250.134(B).

Where required or installed, equipment bonding jumpers shall be installed in accordance with 250.102.

ARTICLE 356
Liquidtight Flexible Nonmetallic Conduit: Type LFNC

I. GENERAL

356.1 Scope. This article covers the use, installation, and construction specifications for liquidtight flexible nonmetallic conduit (LFNC) and associated fittings.

356.2 Definition.

Liquidtight Flexible Nonmetallic Conduit (LFNC). A raceway of circular cross section of various types as follows:

(1) A smooth seamless inner core and cover bonded together and having one or more reinforcement layers between the core and covers, designated as Type LFNC-A

(2) A smooth inner surface with integral reinforcement within the conduit wall, designated as Type LFNC-B

(3) A corrugated internal and external surface without integral reinforcement within the conduit wall, designated as LFNC-C

LFNC is flame resistant and with fittings and is approved for the installation of electrical conductors.

356.6 Listing Requirements. LFNC and associated fittings shall be listed.

II. INSTALLATION

356.10 Uses Permitted. LFNC shall be permitted to be used in exposed or concealed locations for the following purposes:

(1) Where flexibility is required for installation, operation, or maintenance.

(2) Where protection of the contained conductors is required from vapors, liquids, or solids.

(3) For outdoor locations where listed and marked as suitable for the purpose.

(4) For direct burial where listed and marked for the purpose.

(5) Type LFNC-B shall be permitted to be installed in lengths longer than 1.8 m (6 ft) where secured in accordance with 356.30.

(6) Type LFNC-B as a listed manufactured prewired assembly, metric designator 16 through 27 (trade size 1/2 through 1) conduit.

(7) For encasement in concrete where listed for direct burial and installed in compliance with 356.42.

356.12 Uses Not Permitted. LFNC shall not be used as follows:

(1) Where subject to physical damage

(2) Where any combination of ambient and conductor temperatures is in excess of that for which the LFNC is approved

(3) In lengths longer than 1.8 m (6 ft), except as permitted by 356.10(5) or where a longer length is approved as essential for a required degree of flexibility

356.26 Bends—Number in One Run. There shall not be more than the equivalent of four quarter bends (360 degrees total) between pull points, for example, conduit bodies and boxes.

356.28 Trimming. All cut ends of conduit shall be trimmed inside and outside to remove rough edges.

356.30 Securing and Supporting. Type LFNC-B shall be securely fastened and supported in accordance with one of the following:

(1) Where installed in lengths exceeding 1.8 m (6 ft), the conduit shall be securely fastened at intervals not exceeding 900 mm (3 ft) and within 300 mm (12 in.) on each side of every outlet box, junction box, cabinet, or fitting.

(2) Securing or supporting of the conduit shall not be required where it is fished, installed in lengths not exceeding 900 mm (3 ft) at terminals where flexibility is required, or installed in lengths not exceeding 1.8 m (6 ft) from a luminaire terminal connection for tap conductors to luminaires permitted in 410.117(C).

(3) Horizontal runs of LFNC supported by openings through framing members at intervals not exceeding 900 mm (3 ft) and securely fastened within 300 mm (12 in.) of termination points shall be permitted.

(4) Securing or supporting of LFNC-B shall not be required where installed in lengths not exceeding 1.8 m (6 ft) from the last point where the raceway is securely fastened for connections within an accessible ceiling to luminaire(s) or other equipment.

356.42 Couplings and Connectors. Only fittings listed for use with LFNC shall be used. Angle connectors shall not be used for concealed raceway installations. Straight LFNC fittings are permitted for direct burial or encasement in concrete.

356.60 Grounding and Bonding. Where used to connect equipment where flexibility is required, an equipment grounding conductor shall be installed.

Where required or installed, equipment grounding conductors shall be installed in accordance with 250.134(B).

Where required or installed, equipment bonding jumpers shall be installed in accordance with 250.102.

CHAPTER 15

WIRING METHODS AND CONDUCTORS

INTRODUCTION

This chapter contains provisions from Article 300—Wiring Methods and Article 310—Conductors for General Wiring. Article 300 covers wiring methods for all wiring installations unless modified by other articles. This article covers some general installation requirements for both cables and raceways. Some of the topics include: protection against physical damage; underground installations; securing and supporting; and boxes, conduit bodies, or fittings.

Article 310 covers general requirements for conductors, including their ampacity ratings. For dwelling unit applications, conductor ampacities specified in Table 310.15(B)(16) are typically used. Under specified conditions in dwelling units, Table 310.15(B)(7) can be used to size service-entrance conductors, service lateral conductors, and feeder conductors that serve as the main power feeder. Conductors sized by Table 310.15(B)(7) must be fed from a 120/240-volt, 3-wire, single-phase source. These conductors can be installed in a raceway or cable, with or without an equipment grounding conductor.

ARTICLE 300
Wiring Methods

I. GENERAL REQUIREMENTS

300.1 Scope.

(A) All Wiring Installations. This article covers wiring methods for all wiring installations unless modified by other articles.

300.3 Conductors.

(A) Single Conductors. Single conductors specified in Table 310.104(A) shall only be installed where part of a recognized wiring method of Chapter 3.

(B) Conductors of the Same Circuit. All conductors of the same circuit and, where used, the grounded conductor and all equipment grounding conductors and bonding conductors shall be contained within the same raceway, auxiliary gutter, cable tray, cablebus assembly, trench, cable, or cord, unless otherwise permitted in accordance with 300.3(B)(1) through (B)(4).

300.4 Protection Against Physical Damage. Where subject to physical damage, conductors, raceways, and cables shall be protected.

(A) Cables and Raceways Through Wood Members.

(1) Bored Holes. In both exposed and concealed locations, where a cable- or raceway-type wiring method is installed through bored holes in joists, rafters, or wood members, holes shall be bored so that the edge of the hole is not less than 32 mm (1-1/4 in.) from the nearest edge of the wood member. Where this distance cannot be maintained, the cable or raceway shall be protected from penetration by screws or nails by a steel plate(s) or bushing(s), at least 1.6 mm (1/16 in.) thick, and of appropriate length and width installed to cover the area of the wiring.

Exception No. 1: Steel plates shall not be required to protect rigid metal conduit, intermediate metal conduit, rigid nonmetallic conduit, or electrical metallic tubing.

Exception No. 2: A listed and marked steel plate less than 1.6 mm (1/16 in.) thick that provides equal or better protection against nail or screw penetration shall be permitted.

(2) Notches in Wood. Where there is no objection because of weakening the building structure, in both exposed and concealed locations, cables or raceways shall be permitted to be laid in notches in wood studs, joists, rafters, or other wood members where the cable or raceway at those points is protected against nails or screws by a steel plate at least 1.6 mm (1/16 in.) thick, and of appropriate length and width, installed to cover the area of the wiring. The steel plate shall be installed before the building finish is applied.

Exception No. 1: Steel plates shall not be required to protect rigid metal conduit, intermediate metal conduit, rigid nonmetallic conduit, or electrical metallic tubing.

Exception No. 2: A listed and marked steel plate less than 1.6 mm (1/16 in.) thick that provides equal or better protection against nail or screw penetration shall be permitted.

(B) Nonmetallic-Sheathed Cables and Electrical Nonmetallic Tubing Through Metal Framing Members.

(1) Nonmetallic-Sheathed Cable. In both exposed and concealed locations where nonmetallic-sheathed cables pass through either factory- or field-punched, cut, or drilled slots or holes in metal members, the cable shall be protected by listed bushings or listed grommets covering all metal edges that are securely fastened in the opening prior to installation of the cable.

(2) Nonmetallic-Sheathed Cable and Electrical Nonmetallic Tubing. Where nails or screws are likely to penetrate nonmetallic-sheathed cable or electrical nonmetallic tubing, a steel sleeve, steel plate, or steel clip not less than 1.6 mm (1/16 in.) in thickness shall be used to protect the cable or tubing.

Exception: A listed and marked steel plate less than 1.6 mm (1/16 in.) thick that provides equal or better protection against nail or screw penetration shall be permitted.

(C) Cables Through Spaces Behind Panels Designed to Allow Access. Cables or raceway-type wiring methods, installed behind panels designed to allow access, shall be supported according to their applicable articles.

(D) Cables and Raceways Parallel to Framing Members and Furring Strips. In both exposed and concealed locations, where a cable- or raceway-type wiring method is installed parallel to framing members, such as joists, rafters, or studs, or is installed parallel to furring strips, the cable or raceway shall be installed and supported so that the nearest outside surface of the cable or raceway is not less than 32 mm (1-1/4 in.) from the nearest edge of the framing member or furring strips where nails or screws are likely to penetrate. Where this distance cannot be maintained, the cable or raceway shall be protected from penetration by nails or screws by a steel plate, sleeve, or equivalent at least 1.6 mm (1/16 in.) thick.

Exception No. 1: Steel plates, sleeves, or the equivalent shall not be required to protect rigid metal conduit, intermediate metal conduit, rigid nonmetallic conduit, or electrical metallic tubing.

Exception No. 2: For concealed work in finished buildings, or finished panels for prefabricated buildings where such supporting is impracticable, it shall be permissible to fish the cables between access points.

Exception No. 3: A listed and marked steel plate less than 1.6 mm (in.) thick that provides equal or better protection against nail or screw penetration shall be permitted.

(E) Cables, Raceways, or Boxes Installed in or Under Roof Decking. A cable, raceway, or box, installed in exposed or concealed locations under metal-corrugated sheet roof decking, shall be installed and supported so there is not less than 38 mm (1-1/2 in.) measured from the lowest surface of the roof decking to the top of the cable, raceway, or box. A cable, raceway, or box shall not be installed in concealed locations in metal-corrugated, sheet decking–type roof.

Exception: Rigid metal conduit and intermediate metal conduit shall not be required to comply with 300.4(E).

(F) Cables and Raceways Installed in Shallow Grooves. Cable- or raceway-type wiring methods installed in a groove, to be covered by wallboard, siding, paneling, carpeting, or similar finish, shall be protected by 1.6 mm (1/16 in.) thick steel plate, sleeve, or equivalent or by not less than 32-mm (1-1/4-in.) free space for the full length of the groove in which the cable or raceway is installed.

Exception No. 1: Steel plates, sleeves, or the equivalent shall not be required to protect rigid metal conduit, intermediate metal conduit, rigid nonmetallic conduit, or electrical metallic tubing.

Exception No. 2: A listed and marked steel plate less than 1.6 mm (1/16 in.) thick that provides equal or better protection against nail or screw penetration shall be permitted.

(G) Insulated Fittings. Where raceways contain 4 AWG or larger insulated circuit conductors, and these conductors enter a cabinet, a box, an enclosure, or a raceway, the conductors shall be protected by an identified fitting providing a smoothly rounded insulating surface, unless the conductors are separated from the fitting or raceway by identified insulating material that is securely fastened in place.

Exception: Where threaded hubs or bosses that are an integral part of a cabinet, box, enclosure, or raceway provide a smoothly rounded or flared entry for conductors.

Conduit bushings constructed wholly of insulating material shall not be used to secure a fitting or raceway. The insulating fitting or insulating material shall have a temperature rating not less than the insulation temperature rating of the installed conductors.

(H) Structural Joints. A listed expansion/deflection fitting or other approved means shall be used where a raceway crosses a structural joint intended for expansion, contraction or deflection, used in buildings, bridges, parking garages, or other structures.

300.5 Underground Installations.

(A) Minimum Cover Requirements. Direct-buried cable or conduit or other raceways shall be installed to meet the minimum cover requirements of Table 300.5.

Table 300.5 Minimum Cover Requirements, 0 to 600 Volts, Nominal, Burial in Millimeters (Inches)

Type of Wiring Method or Circuit

Location of Wiring Method or Circuit	Column 1 Direct Burial Cables or Conductors		Column 2 Rigid Metal Conduit or Intermediate Metal Conduit		Column 3 Nonmetallic Raceways Listed for Direct Burial Without Concrete Encasement or Other Approved Raceways		Column 4 Residential Branch Circuits Rated 120 Volts or Less with GFCI Protection and Maximum Overcurrent Protection of 20 Amperes		Column 5 Circuits for Control of Irrigation and Landscape Lighting Limited to Not More Than 30 Volts and Installed with Type UF or in Other Identified Cable or Raceway	
	mm	in.	mm	in.	mm	in.	mm	in.	mm	in.
All locations not specified below	600	24	150	6	450	18	300	12	150	6
In trench below 50-mm (2-in.) thick concrete or equivalent	450	18	150	6	300	12	150	6	150	6
Under a building	0 (in raceway or Type MC or Type MI cable identified for direct burial)	0	0	0	0 (in raceway or Type MC or Type MI cable identified for direct burial)	0	0 (in raceway or Type MC or Type MI cable identified for direct burial)	0	0 (in raceway or Type MC or Type MI cable identified for direct burial)	0

Location	(mm)	(in.)	(mm)	(in.)	(mm)	(in.)	(mm)	(in.)	(mm)	(in.)
Under minimum of 102-mm (4-in.) thick concrete exterior slab with no vehicular traffic and the slab extending not less than 152 mm (6 in.) beyond the underground installation	450	18	100	4	100	4	150 (direct burial) 100 (in raceway)	6 (direct burial) 4 (in raceway)	150 (direct burial) 100 (in raceway)	6 (direct burial) 4 (in raceway)
Under streets, highways, roads, alleys, driveways, and parking lots	600	24	600	24	600	24	600	24	600	24
One- and two-family dwelling driveways and outdoor parking areas, and used only for dwelling-related purposes	450	18	450	18	450	18	300	12	450	18
In or under airport runways, including adjacent areas where trespassing prohibited	450	18	450	18	450	18	450	18	450	18

(continued)

Table 300.5 Minimum Cover Requirements, 0 to 600 Volts, Nominal, Burial in Millimeters (Inches) *Continued*

Notes:

1. Cover is defined as the shortest distance in millimeters (inches) measured between a point on the top surface of any direct-buried conductor, cable, conduit, or other raceway and the top surface of finished grade, concrete, or similar cover.

2. Raceways approved for burial only where concrete encased shall require concrete envelope not less than 50 mm (2 in.) thick.

3. Lesser depths shall be permitted where cables and conductors rise for terminations or splices or where access is otherwise required.

4. Where one of the wiring method types listed in Columns 1–3 is used for one of the circuit types in Columns 4 and 5, the shallowest depth of burial shall be permitted.

5. Where solid rock prevents compliance with the cover depths specified in this table, the wiring shall be installed in metal or non-metallic raceway permitted for direct burial. The raceways shall be covered by a minimum of 50 mm (2 in.) of concrete extending down to rock.

(B) Wet Locations. The interior of enclosures or raceways installed underground shall be considered to be a wet location. Insulated conductors and cables installed in these enclosures or raceways in underground installations shall be listed for use in wet locations and shall comply with 310.10(C). Any connections or splices in an underground installation shall be approved for wet locations.

(C) Underground Cables Under Buildings. Underground cable installed under a building shall be in a raceway.

Exception No. 2: Type MC Cable listed for direct burial or concrete encasement shall be permitted under a building without installation in a raceway in accordance with 330.10(A)(5) and in wet locations in accordance with 330.10(11).

(D) Protection from Damage. Direct-buried conductors and cables shall be protected from damage in accordance with 300.5(D)(1) through (D)(4).

(1) Emerging from Grade. Direct-buried conductors and cables emerging from grade and specified in columns 1 and 4 of Table 300.5 shall be protected by enclosures or raceways extending from the minimum cover distance below grade required by 300.5(A) to a point at least 2.5 m (8 ft) above finished grade. In no case shall the protection be required to exceed 450 mm (18 in.) below finished grade.

(2) Conductors Entering Buildings. Conductors entering a building shall be protected to the point of entrance.

(3) Service Conductors. Underground service conductors that are not encased in concrete and that are buried 450 mm (18 in.) or more below grade shall have their location identified by a warning ribbon that is placed in the trench at least 300 mm (12 in.) above the underground installation.

(4) Enclosure or Raceway Damage. Where the enclosure or raceway is subject to physical damage, the conductors shall be installed in rigid metal conduit, intermediate metal conduit, Schedule 80 PVC conduit, or equivalent.

(E) Splices and Taps. Direct-buried conductors or cables shall be permitted to be spliced or tapped without the use of splice boxes. The splices or taps shall be made in accordance with 110.14(B).

(F) Backfill. Backfill that contains large rocks, paving materials, cinders, large or sharply angular substances, or corrosive material shall not be placed in an excavation where materials may damage raceways, cables, or other substructures or prevent adequate compaction of fill or contribute to corrosion of raceways, cables, or other substructures.

Where necessary to prevent physical damage to the raceway or cable, protection shall be provided in the form of granular or selected material, suitable running boards, suitable sleeves, or other approved means.

(G) Raceway Seals. Conduits or raceways through which moisture may contact live parts shall be sealed or plugged at either or both ends.

(H) Bushing. A bushing, or terminal fitting, with an integral bushed opening shall be used at the end of a conduit or other raceway that terminates underground where the conductors or cables emerge as a direct burial wiring method. A seal incorporating the physical protection characteristics of a bushing shall be permitted to be used in lieu of a bushing.

(I) Conductors of the Same Circuit. All conductors of the same circuit and, where used, the grounded conductor and all equipment grounding conductors shall be installed in the same raceway or cable or shall be installed in close proximity in the same trench.

Exception No. 1: Conductors shall be permitted to be installed in parallel in raceways, multiconductor cables, or direct-buried single conductor cables. Each raceway or multiconductor cable shall contain all conductors of the same circuit, including equipment grounding conductors. Each direct-buried single conductor cable shall be located in close proximity in the trench to the other single conductor cables in the same parallel set of conductors in the circuit, including equipment grounding conductors.

(J) Earth Movement. Where direct-buried conductors, raceways, or cables are subject to movement by settlement or frost, direct-buried conductors, raceways, or cables shall be arranged so as to prevent damage to the enclosed conductors or to equipment connected to the raceways.

(K) Directional Boring. Cables or raceways installed using directional boring equipment shall be approved for the purpose.

300.9 Raceways in Wet Locations Aboveground. Where raceways are installed in wet locations aboveground, the interior of these raceways shall be considered to be a wet location. Insulated conductors and cables installed in raceways in wet locations aboveground shall comply with 310.10(C).

300.10 Electrical Continuity of Metal Raceways and Enclosures. Metal raceways, cable armor, and other metal enclosures for conductors shall be metallically joined together into a continuous electrical conductor and shall be connected to all boxes, fittings, and cabinets so as to provide effective electrical continuity. Unless specifically permitted elsewhere in this *Code,* raceways and cable assemblies shall be mechanically secured to boxes, fittings, cabinets, and other enclosures.

Exception No. 1: Short sections of raceways used to provide support or protection of cable assemblies from physical damage shall not be required to be made electrically continuous.

300.11 Securing and Supporting.

(A) Secured in Place. Raceways, cable assemblies, boxes, cabinets, and fittings shall be securely fastened in place. Support wires that do not provide secure support shall not be permitted as the sole support. Support wires and associated fittings that provide secure support and that are installed in addition to the ceiling grid support wires shall be permitted as the sole support. Where independent support wires are used, they shall be secured at both ends. Cables and raceways shall not be supported by ceiling grids.

(1) Fire-Rated Assemblies. Wiring located within the cavity of a fire-rated floor–ceiling or roof–ceiling assembly shall not be secured to, or supported by, the ceiling assembly, including the ceiling support wires. An independent means of secure support shall be provided and shall be permitted to be attached to the assembly. Where independent support wires are used, they shall be distinguishable by color, tagging, or other effective means from those that are part of the fire-rated design.

Exception: The ceiling support system shall be permitted to support wiring and equipment that have been tested as part of the fire-rated assembly.

(2) Non–Fire-Rated Assemblies. Wiring located within the cavity of a non–fire-rated floor–ceiling or roof–ceiling assembly shall not be secured to, or supported by, the ceiling assembly, including the ceiling support wires. An independent means of secure support shall be provided and shall be permitted to be attached to the assembly. Where independent support wires are used, they shall be distinguishable by color, tagging, or other effective means.

Exception: The ceiling support system shall be permitted to support branch-circuit wiring and associated equipment where installed in accordance with the ceiling system manufacturer's instructions.

(B) Raceways Used as Means of Support. Raceways shall be used only as a means of support for other raceways, cables, or nonelectrical equipment under any of the following conditions:

(1) Where the raceway or means of support is identified for the purpose

(2) Where the raceway contains power supply conductors for electrically controlled equipment and is used to support Class 2 circuit conductors or cables that are solely for the purpose of connection to the equipment control circuits

(3) Where the raceway is used to support boxes or conduit bodies in accordance with 314.23 or to support luminaires in accordance with 410.36(E)

(C) Cables Not Used as Means of Support. Cable wiring methods shall not be used as a means of support for other cables, raceways, or nonelectrical equipment.

300.12 Mechanical Continuity—Raceways and Cables. Metal or nonmetallic raceways, cable armors, and cable sheaths shall be continuous between cabinets, boxes, fittings, or other enclosures or outlets.

Exception No. 1: Short sections of raceways used to provide support or protection of cable assemblies from physical damage shall not be required to be mechanically continuous.

300.13 Mechanical and Electrical Continuity—Conductors.

(A) General. Conductors in raceways shall be continuous between outlets, boxes, devices, and so forth. There shall be no splice or tap within a raceway unless permitted by 300.15; 368.56(A); 376.56; 378.56; 384.56; 386.56; 388.56; or 390.7.

(B) Device Removal. In multiwire branch circuits, the continuity of a grounded conductor shall not depend on device connections such as lampholders, receptacles, and so forth, where the removal of such devices would interrupt the continuity.

300.14 Length of Free Conductors at Outlets, Junctions, and Switch Points.
At least 150 mm (6 in.) of free conductor, measured from the point in the box where it emerges from its raceway or cable sheath, shall be left at each outlet, junction, and switch point for splices or the connection of luminaires or devices. Where the opening to an outlet, junction, or switch point is less than 200 mm (8 in.) in any dimension, each conductor shall be long enough to extend at least 75 mm (3 in.) outside the opening.

Exception: Conductors that are not spliced or terminated at the outlet, junction, or switch point shall not be required to comply with 300.14.

300.15 Boxes, Conduit Bodies, or Fittings—Where Required.
A box shall be installed at each outlet and switch point for concealed knob-and-tube wiring.

Fittings and connectors shall be used only with the specific wiring methods for which they are designed and listed.

Where the wiring method is conduit, tubing, Type AC cable, Type MC cable, Type MI cable, nonmetallic-sheathed cable, or other cables, a box or conduit body shall be installed at each conductor splice point, outlet point, switch point, junction point, termination point, or pull point, unless otherwise permitted in 300.15(A) through (L).

(A) Wiring Methods with Interior Access. A box or conduit body shall not be required for each splice, junction, switch, pull, termination, or outlet points in wiring methods with removable covers, such as wireways, multioutlet assemblies, auxiliary gutters, and surface raceways. The covers shall be accessible after installation.

(B) Equipment. An integral junction box or wiring compartment as part of approved equipment shall be permitted in lieu of a box.

(C) Protection. A box or conduit body shall not be required where cables enter or exit from conduit or tubing that is used to provide cable support or protection against physical damage. A fitting shall be provided on the end(s) of the conduit or tubing to protect the cable from abrasion.

(E) Integral Enclosure. A wiring device with integral enclosure identified for the use, having brackets that securely fasten the device to walls or ceilings of conventional on-site frame construction, for use with nonmetallic-sheathed cable, shall be permitted in lieu of a box or conduit body.

(F) Fitting. A fitting identified for the use shall be permitted in lieu of a box or conduit body where conductors are not spliced or terminated within the fitting. The fitting shall be accessible after installation.

(G) Direct-Buried Conductors. As permitted in 300.5(E), a box or conduit body shall not be required for splices and taps in direct-buried conductors and cables.

(H) Insulated Devices. As permitted in 334.40(B), a box or conduit body shall not be required for insulated devices supplied by nonmetallic-sheathed cable.

(J) Luminaires. A box or conduit body shall not be required where a luminaire is used as a raceway as permitted in 410.64.

300.16 Raceway or Cable to Open or Concealed Wiring.

(A) Box, Conduit Body, or Fitting. A box, conduit body, or terminal fitting having a separately bushed hole for each conductor shall be used wherever a change is made from conduit, electrical metallic tubing, electrical nonmetallic tubing, nonmetallic-sheathed cable, Type AC cable, Type MC cable, or mineral-insulated, metal-sheathed cable and surface raceway wiring to open wiring or to concealed knob-and-tube wiring. A fitting used for this purpose shall contain no taps or splices and shall not be used at luminaire outlets. A conduit body used for this purpose shall contain no taps or splices, unless it complies with 314.16(C)(2).

(B) Bushing. A bushing shall be permitted in lieu of a box or terminal where the conductors emerge from a raceway and enter or terminate at equipment, such as open switchboards, unenclosed control equipment, or similar equipment. The bushing shall be of the insulating type for other than lead-sheathed conductors.

300.18 Raceway Installations.

(A) Complete Runs. Raceways, other than busways or exposed raceways having hinged or removable covers, shall be installed complete between outlet, junction, or splicing points prior to the installation of conductors. Where required to facilitate the installation of utilization equipment, the raceway shall be permitted to be initially installed without a terminating connection at the equipment. Prewired raceway assemblies shall be permitted only where specifically permitted in this *Code* for the applicable wiring method.

Exception: Short sections of raceways used to contain conductors or cable assemblies for protection from physical damage shall not be required to be installed complete between outlet, junction, or splicing points.

300.20 Induced Currents in Ferrous Metal Enclosures or Ferrous Metal Raceways.

(A) Conductors Grouped Together. Where conductors carrying alternating current are installed in ferrous metal enclosures or ferrous metal raceways, they shall be arranged so as to avoid heating the surrounding ferrous metal by induction. To accomplish this, all phase conductors and, where used, the grounded conductor and all equipment grounding conductors shall be grouped together.

300.21 Spread of Fire or Products of Combustion. Electrical installations in hollow spaces, vertical shafts, and ventilation or air-handling ducts shall be made so that the possible spread of fire or products of combustion will not be substantially increased. Openings around electrical penetrations into or through fire-resistant-rated walls, partitions, floors, or ceilings shall be firestopped using approved methods to maintain the fire resistance rating.

300.22 Wiring in Ducts Not Used for Air Handling, Fabricated Ducts for Environmental Air, and Other Spaces for Environmental Air (Plenums). The provisions of this section shall apply to the installation and uses of electrical wiring and equipment in ducts used for dust, loose stock, or vapor removal; ducts specifically fabricated for environmental air; and other spaces used for environmental air (plenums).

(C) Other Spaces Used for Environmental Air (Plenums). This section shall apply to spaces not specifically fabricated for environmental air-handling purposes but used for air-handling purposes as a plenum. This section shall not apply to habitable rooms or areas of buildings, the prime purpose of which is not air handling.

Exception: This section shall not apply to the joist or stud spaces of dwelling units where the wiring passes through such spaces perpendicular to the long dimension of such spaces.

ARTICLE 310
Conductors for General Wiring

I. GENERAL

310.1 Scope. This article covers general requirements for conductors and their type designations, insulations, markings, mechanical strengths, ampacity ratings, and uses. These requirements do not apply to conductors that form an integral part of equipment, such as motors, motor controllers, and similar equipment, or to conductors specifically provided for elsewhere in this *Code*.

II. INSTALLATION

310.10 Uses Permitted. The conductors described in 310.104 shall be permitted for use in any of the wiring methods covered in Chapter 3 and as specified in their respective tables or as permitted elsewhere in this *Code*.

(A) Dry Locations. Insulated conductors and cables used in dry locations shall be any of the types identified in this *Code*.

(B) Dry and Damp Locations. Insulated conductors and cables used in dry and damp locations shall be Types FEP, FEPB, MTW, PFA, RHH, RHW, RHW-2, SA, THHN, THW, THW-2, THHW, THWN, THWN-2, TW, XHH, XHHW, XHHW-2, Z, or ZW.

(C) Wet Locations. Insulated conductors and cables used in wet locations shall comply with one of the following:

(1) Be moisture-impervious metal-sheathed
(2) Be types MTW, RHW, RHW-2, TW, THW, THW-2, THHW, THWN, THWN-2, XHHW, XHHW-2, ZW
(3) Be of a type listed for use in wet locations

(D) Locations Exposed to Direct Sunlight. Insulated conductors or cables used where exposed to direct rays of the sun shall comply with (D)(1) or (D)(2):

(1) Conductors and cables shall be listed, or listed and marked, as being sunlight resistant
(2) Conductors and cables shall be covered with insulating material, such as tape or sleeving, that is listed, or listed and marked, as being sunlight resistant

(F) Direct-Burial Conductors. Conductors used for direct-burial applications shall be of a type identified for such use.

(H) Conductors in Parallel.

(1) General. Aluminum, copper-clad aluminum, or copper conductors, for each phase, polarity, neutral, or grounded circuit shall be permitted to be connected in parallel (electrically joined at both ends) only in sizes 1/0 AWG and larger where installed in accordance with 310.10(H)(2) through (H)(6).

(2) Conductor Characteristics. The paralleled conductors in each phase, polarity, neutral, grounded circuit conductor, equipment grounding conductor, or equipment bonding jumper shall comply with all of the following:

(1) Be the same length
(2) Consist of the same conductor material
(3) Be the same size in circular mil area
(4) Have the same insulation type
(5) Be terminated in the same manner

(3) Separate Cables or Raceways. Where run in separate cables or raceways, the cables or raceways with conductors shall have the same number of conductors and shall have the same electrical characteristics. Conductors of one phase, polarity, neutral, grounded circuit conductor, or equipment grounding conductor shall not be required to have the same physical characteristics as those of another phase, polarity, neutral, grounded circuit conductor, or equipment grounding conductor.

(4) Ampacity Adjustment. Conductors installed in parallel shall comply with the provisions of 310.15(B)(3)(a).

(5) Equipment Grounding Conductors. Where parallel equipment grounding conductors are used, they shall be sized in accordance with 250.122. Sectioned equipment grounding conductors smaller than 1/0 AWG shall be permitted in multiconductor cables in accordance with 310.104, provided the combined circular mil area of the sectioned equipment grounding conductors in each cable complies with 250.122.

(6) Equipment Bonding Jumpers. Where parallel equipment bonding jumpers are installed in raceways, they shall be sized and installed in accordance with 250.102.

310.15 Ampacities for Conductors Rated 0–2000 Volts.

(B) Tables. Ampacities for conductors rated 0 to 2000 volts shall be as specified in the Allowable Ampacity Table 310.15(B)(16) through Table 310.15(B)(19), and Ampacity Table 310.15(B)(20) and Table 310.15(B)(21) as modified by 310.15(B)(1) through (B)(7).

The temperature correction and adjustment factors shall be permitted to be applied to the ampacity for the temperature rating of the conductor, if the corrected and adjusted ampacity does not exceed the ampacity for the temperature rating of the termination in accordance with the provisions of 110.14(C).

Informational Note: Table 310.15(B)(16) through Table 310.15(B)(19) are application tables for use in determining conductor sizes on loads calculated in accordance with Article 220. Allowable ampacities result from consideration of one or more of the following:

(1) Temperature compatibility with connected equipment, especially the connection points.
(2) Coordination with circuit and system overcurrent protection.
(3) Compliance with the requirements of product listings or certifications. See 110.3(B).
(4) Preservation of the safety benefits of established industry practices and standardized procedures.

(2) Ambient Temperature Correction Factors. Ampacities for ambient temperatures other than those shown in the ampacity tables shall be corrected in accordance with Table 310.15(B)(2)(a).

(3) Adjustment Factors.

(a) *More Than Three Current-Carrying Conductors in a Raceway or Cable.* Where the number of current-carrying conductors in a raceway or cable exceeds three, or where single conductors or multiconductor cables are installed without maintaining spacing for a continuous length longer than 600 mm (24 in.) and are not installed in raceways, the allowable ampacity of each conductor shall be reduced as shown in Table 310.15(B)(3)(a). Each current-carrying conductor of a paralleled set of conductors shall be counted as a current-carrying conductor.

(2) Adjustment factors shall not apply to conductors in raceways having a length not exceeding 600 mm (24 in.).
(3) Adjustment factors shall not apply to underground conductors entering or leaving an outdoor trench if those

Table 310.15(B)(2)(a) Ambient Temperature Correction Factors Based on 30°C (86°F)

For ambient temperatures other than 30°C (86°F), multiply the allowable ampacities specified in the ampacity tables by the appropriate correction factor shown below.

Ambient Temperature (°C)	Temperature Rating of Conductor			Ambient Temperature (°F)
	60°C	75°C	90°C	
10 or less	1.29	1.20	1.15	50 or less
11–15	1.22	1.15	1.12	51–59
16–20	1.15	1.11	1.08	60–68
21–25	1.08	1.05	1.04	69–77
26–30	1.00	1.00	1.00	78–86
31–35	0.91	0.94	0.96	87–95
36–40	0.82	0.88	0.91	96–104
41–45	0.71	0.82	0.87	105–113
46–50	0.58	0.75	0.82	114–122
51–55	0.41	0.67	0.76	123–131
56–60	—	0.58	0.71	132–140
61–65	—	0.47	0.65	141–149
66–70	—	0.33	0.58	150–158
71–75	—	—	0.50	159–167
76–80	—	—	0.41	168–176
81–85	—	—	0.29	177–185

conductors have physical protection in the form of rigid metal conduit, intermediate metal conduit, rigid polyvinyl chloride conduit (PVC), or reinforced thermosetting resin conduit (RTRC) having a length not exceeding 3.05 m (10 ft), and if the number of conductors does not exceed four.

(4) Bare or Covered Conductors. Where bare or covered conductors are installed with insulated conductors, the temperature rating of the bare or covered conductor shall be equal to the lowest temperature rating of the insulated conductors for the purpose of determining ampacity.

Table 310.15(B)(3)(a) Adjustment Factors for More Than Three Current-Carrying Conductors in a Raceway or Cable

Number of Conductors[1]	Percent of Values in Table 310.15(B)(16) through Table 310.15(B)(19) as Adjusted for Ambient Temperature if Necessary
4–6	80
7–9	70
10–20	50
21–30	45
31–40	40
41 and above	35

[1]Number of conductors is the total number of conductors in the raceway or cable adjusted in accordance with 310.15(B)(5) and (6).

(5) Neutral Conductor.

(a) A neutral conductor that carries only the unbalanced current from other conductors of the same circuit shall not be required to be counted when applying the provisions of 310.15(B)(3)(a).

(6) Grounding or Bonding Conductor. A grounding or bonding conductor shall not be counted when applying the provisions of 310.15(B)(3)(a).

(7) 120/240-Volt, 3-Wire, Single-Phase Dwelling Services and Feeders. For individual dwelling units of one-family, two-family, and multifamily dwellings, conductors, as listed in Table 310.15(B)(7), shall be permitted as 120/240-volt, 3-wire, single-phase service-entrance conductors, service-lateral conductors, and feeder conductors that serve as the main power feeder to each dwelling unit and are installed in raceway or cable with or without an equipment grounding conductor. For application of this section, the main power feeder shall be the feeder between the main disconnect and the panelboard that supplies, either by branch circuits or by feeders, or both, all loads that are part or associated with the dwelling unit. The feeder conductors to a dwelling unit shall not be required to have an allowable ampacity rating greater

Table 310.15(B)(7) Conductor Types and Sizes for 120/240-Volt, 3-Wire, Single-Phase Dwelling Services and Feeders. Conductor Types RHH, RHW, RHW-2, THHN, THHW, THW, THW-2, THWN, THWN-2, XHHW, XHHW-2, SE, USE, USE-2

| Service or Feeder Rating (Amperes) | Conductor (AWG or kcmil) | |
	Copper	Aluminum or Copper-Clad Aluminum
100	4	2
110	3	1
125	2	1/0
150	1	2/0
175	1/0	3/0
200	2/0	4/0
225	3/0	250
250	4/0	300
300	250	350
350	350	500
400	400	600

than their service-entrance conductors. The grounded conductor shall be permitted to be smaller than the ungrounded conductors, provided the requirements of 215.2, 220.61, and 230.42 are met.

III. CONSTRUCTION SPECIFICATIONS

310.106 Conductors.

(B) Conductor Material. Conductors in this article shall be of aluminum, copper-clad aluminum, or copper unless otherwise specified.

(C) Stranded Conductors. Where installed in raceways, conductors 8 AWG and larger, not specifically permitted or required elsewhere in this *Code* to be solid, shall be stranded.

(D) Insulated. Conductors, not specifically permitted elsewhere in this *Code* to be covered or bare, shall be insulated.

Table 310.15(B)(16) (formerly Table 310.16) Allowable Ampacities of Insulated Conductors Rated Up to and Including 2000 Volts, 60°C Through 90°C (140°F Through 194°F), Not More Than Three Current-Carrying Conductors in Raceway, Cable, or Earth (Directly Buried), Based on Ambient Temperature of 30°C (86°F)*

Size AWG or kcmil	Temperature Rating of Conductor [See Table 310.104(A).]						Size AWG or kcmil
	60°C (140°F)	75°C (167°F)	90°C (194°F)	60°C (140°F)	75°C (167°F)	90°C (194°F)	
	Types TW, UF	Types RHW, THHW, THW, THWN, XHHW, USE, ZW	Types TBS, SA, SIS, FEP, FEPB, MI, RHH, RHW-2, THHN, THHW, THW-2, THWN-2, USE-2, XHH, XHHW, XHHW-2, ZW-2	Types TW, UF	Types RHW, THHW, THW, THWN, XHHW, USE	Types TBS, THHW, THW-2, THWN-2, RHH, RHW-2, THHN, THWN-2, USE-2, XHH, XHHW, XHHW-2, ZW-2	
	COPPER			ALUMINUM OR COPPER-CLAD ALUMINUM			
18	—	—	14				
16	—	—	18				—
14**	15	20	25	—	—	—	—
12**	20	25	30	15	20	25	12**
10**	30	35	40	25	30	35	10**
8	40	50	55	35	40	45	8

(continued)

Table 310.15(B)(16) (formerly Table 310.16) Allowable Ampacities of Insulated Conductors Rated Up to and Including 2000 Volts, 60°C Through 90°C (140°F Through 194°F), Not More Than Three Current-Carrying Conductors in Raceway, Cable, or Earth (Directly Buried), Based on Ambient Temperature of 30°C (86°F).* *Continued*

Size AWG or kcmil	Temperature Rating of Conductor [See Table 310.104(A).]						Size AWG or kcmil
	60°C (140°F)	75°C (167°F)	90°C (194°F)	60°C (140°F)	75°C (167°F)	90°C (194°F)	
	Types TW, UF	Types RHW, THW, THHW, THWN, XHHW, USE, ZW	Types TBS, SA, SIS, FEP, FEPB, MI, RHH, RHW-2, THHN, THHW, THW-2, THWN-2, USE-2, XHH, XHHW, XHHW-2, ZW-2	Types TW, UF	Types RHW, THW, THHW, THWN, XHHW, USE	Types TBS, SA, SIS, THHN, THHW, THW-2, THWN-2, RHH, RHW-2, USE-2, XHH, XHHW, XHHW-2, ZW-2	
	COPPER			ALUMINUM OR COPPER-CLAD ALUMINUM			
6	55	65	75	40	50	55	6
4	70	85	95	55	65	75	4
3	85	100	115	65	75	85	3
2	95	115	130	75	90	100	2
1	110	130	145	85	100	115	1
1/0	125	150	170	100	120	135	1/0
2/0	145	175	195	115	135	150	2/0

Size	175 / 205	155 / 180	130 / 150	225 / 260	200 / 230	165 / 195	Size
3/0	175	155	130	225	200	165	3/0
4/0	205	180	150	260	230	195	4/0
250	230	205	170	290	255	215	250
300	260	230	195	320	285	240	300
350	280	250	210	350	310	260	350
400	305	270	225	380	335	280	400
500	350	310	260	430	380	320	500
600	385	340	285	475	420	350	600
700	425	375	315	520	460	385	700
750	435	385	320	535	475	400	750
800	445	395	330	555	490	410	800
900	480	425	355	585	520	435	900
1000	500	445	375	615	545	455	1000
1250	545	485	405	665	590	495	1250
1500	585	520	435	705	625	525	1500
1750	615	545	455	735	650	545	1750
2000	630	560	470	750	665	555	2000

*Refer to 310.15(B)(2) for the ampacity correction factors where the ambient temperature is other than 30°C (86°F).

**Refer to 240.4(D) for conductor overcurrent protection limitations.

CHAPTER 16

BRANCH CIRCUITS

INTRODUCTION

This chapter contains provisions from Article 210—Branch Circuits. The provisions within Article 210 have been divided into three chapters of this pocket guide. This chapter covers Part I (general provisions) and Part II (branch-circuit ratings). Some of the general provisions pertain to multiwire branch circuits, ground-fault circuit-interrupters, required branch circuits, and arc-fault circuit interrupters. Part II covers the branch-circuit ratings for conductors and receptacles.

Chapter 17 contains the receptacle outlet provisions within Article 210, and Chapter 19 contains the lighting outlet provisions.

ARTICLE 210
Branch Circuits

I. GENERAL PROVISIONS

210.1 Scope. This article covers branch circuits except for branch circuits that supply only motor loads, which are covered in Article 430. Provisions of this article and Article 430 apply to branch circuits with combination loads.

210.3 Rating. Branch circuits recognized by this article shall be rated in accordance with the maximum permitted ampere rating or setting of the overcurrent device. The rating for other than individual branch circuits shall be 15, 20, 30, 40, and 50 amperes. Where conductors of higher ampacity are used for any reason, the ampere rating or setting of the specified overcurrent device shall determine the circuit rating.

210.4 Multiwire Branch Circuits.

(A) General. Branch circuits recognized by this article shall be permitted as multiwire circuits. A multiwire circuit shall be permitted to be considered as multiple circuits. All conductors of a multiwire branch circuit shall originate from the same panelboard or similar distribution equipment.

(B) Disconnecting Means. Each multiwire branch circuit shall be provided with a means that will simultaneously disconnect all ungrounded conductors at the point where the branch circuit originates.

(C) Line-to-Neutral Loads. Multiwire branch circuits shall supply only line-to-neutral loads.

Exception No. 1: A multiwire branch circuit that supplies only one utilization equipment.

Exception No. 2: Where all ungrounded conductors of the multiwire branch circuit are opened simultaneously by the branch-circuit overcurrent device.

(D) Grouping. The ungrounded and grounded circuit conductors of each multiwire branch circuit shall be grouped by cable ties or similar means in at least one location within the panelboard or other point of origination.

Exception: The requirement for grouping shall not apply if the circuit enters from a cable or raceway unique to the circuit that makes the grouping obvious.

210.5 Identification for Branch Circuits.

(A) Grounded Conductor. The grounded conductor of a branch circuit shall be identified in accordance with 200.6.

(B) Equipment Grounding Conductor. The equipment grounding conductor shall be identified in accordance with 250.119.

210.7 Multiple Branch Circuits.
Where two or more branch circuits supply devices or equipment on the same yoke, a means to simultaneously disconnect the ungrounded conductors supplying those devices shall be provided at the point at which the branch circuits originate.

210.8 Ground-Fault Circuit-Interrupter Protection for Personnel.
Ground-fault circuit-interruption for personnel shall be provided as required in 210.8(A) through (C). The ground-fault circuit-interrupter shall be installed in a readily accessible location.

(A) Dwelling Units. All 125-volt, single-phase, 15- and 20-ampere receptacles installed in the locations specified in 210.8(A)(1) through (8) shall have ground-fault circuit-interrupter protection for personnel.

(1) Bathrooms
(2) Garages, and also accessory buildings that have a floor located at or below grade level not intended as habitable rooms and limited to storage areas, work areas, and areas of similar use
(3) Outdoors

Exception to (3): Receptacles that are not readily accessible and are supplied by a branch circuit dedicated to electric snow-melting, deicing, or pipeline and vessel heating equipment shall be permitted to be installed in accordance with 426.28 or 427.22, as applicable.

(4) Crawl spaces—at or below grade level
(5) Unfinished basements—for purposes of this section, unfinished basements are defined as portions or areas

of the basement not intended as habitable rooms and limited to storage areas, work areas, and the like

Exception to (5): A receptacle supplying only a permanently installed fire alarm or burglar alarm system shall not be required to have ground-fault circuit-interrupter protection.

Receptacles installed under the exception to 210.8(A)(5) shall not be considered as meeting the requirements of 210.52(G).

(6) Kitchens—where the receptacles are installed to serve the countertop surfaces
(7) Sinks—located in areas other than kitchens where receptacles are installed within 1.8 m (6 ft) of the outside edge of the sink
(8) Boathouses

(C) Boat Hoists. GFCI protection shall be provided for outlets not exceeding 240 volts that supply boat hoists installed in dwelling unit locations.

210.11 Branch Circuits Required. Branch circuits for lighting and for appliances, including motor-operated appliances, shall be provided to supply the loads calculated in accordance with 220.10. In addition, branch circuits shall be provided for specific loads not covered by 220.10 where required elsewhere in this *Code* and for dwelling unit loads as specified in 210.11(C).

(A) Number of Branch Circuits. The minimum number of branch circuits shall be determined from the total calculated load and the size or rating of the circuits used. In all installations, the number of circuits shall be sufficient to supply the load served. In no case shall the load on any circuit exceed the maximum specified by 220.18.

(B) Load Evenly Proportioned Among Branch Circuits. Where the load is calculated on the basis of volt-amperes per square meter or per square foot, the wiring system up to and including the branch-circuit panelboard(s) shall be provided to serve not less than the calculated load. This load shall be evenly proportioned among multioutlet branch circuits within the panelboard(s). Branch-circuit overcurrent devices and circuits shall be required to be installed only to serve the connected load.

(C) Dwelling Units.

(1) Small-Appliance Branch Circuits. In addition to the number of branch circuits required by other parts of this section, two or more 20-ampere small-appliance branch circuits shall be provided for all receptacle outlets specified by 210.52(B).

(2) Laundry Branch Circuits. In addition to the number of branch circuits required by other parts of this section, at least one additional 20-ampere branch circuit shall be provided to supply the laundry receptacle outlet(s) required by 210.52(F). This circuit shall have no other outlets.

(3) Bathroom Branch Circuits. In addition to the number of branch circuits required by other parts of this section, at least one 20-ampere branch circuit shall be provided to supply bathroom receptacle outlet(s). Such circuits shall have no other outlets.

Exception: Where the 20-ampere circuit supplies a single bathroom, outlets for other equipment within the same bathroom shall be permitted to be supplied in accordance with 210.23(A)(1) and (A)(2).

210.12 Arc-Fault Circuit-Interrupter Protection.

(A) Dwelling Units. All 120-volt, single phase, 15- and 20-ampere branch circuits supplying outlets installed in dwelling unit family rooms, dining rooms, living rooms, parlors, libraries, dens, bedrooms, sunrooms, recreation rooms, closets, hallways, or similar rooms or areas shall be protected by a listed arc-fault circuit interrupter, combination-type, installed to provide protection of the branch circuit.

Exception No. 1: If RMC, IMC, EMT, Type MC, or steel armored Type AC cables meeting the requirements of 250.118 and metal outlet and junction boxes are installed for the portion of the branch circuit between the branch-circuit overcurrent device and the first outlet, it shall be permitted to install an outlet branch-circuit type AFCI at the first outlet to provide protection for the remaining portion of the branch circuit.

Exception No. 2: Where a listed metal or nonmetallic conduit or tubing is encased in not less than 50 mm (2 in.) of concrete for the portion of the branch circuit between the branch-circuit overcurrent device and the first outlet, it shall be permitted to install an outlet branch-circuit type AFCI at the first outlet to provide protection for the remaining portion of the branch circuit.

Exception No. 3: Where an individual branch circuit to a fire alarm system installed in accordance with 760.41(B) or 760.121(B) is installed in RMC, IMC, EMT, or steel-sheathed cable, Type AC or Type MC, meeting the requirements of 250.118, with metal outlet and junction boxes, AFCI protection shall be permitted to be omitted.

(B) Branch Circuit Extensions or Modifications— Dwelling Units. In any of the areas specified in 210.12(A), where branch-circuit wiring is modified, replaced, or extended, the branch circuit shall be protected by one of the following:

(1) A listed combination-type AFCI located at the origin of the branch circuit
(2) A listed outlet branch-circuit type AFCI located at the first receptacle outlet of the existing branch circuit

II. BRANCH-CIRCUIT RATINGS

210.19 Conductors—Minimum Ampacity and Size.

(A) Branch Circuits Not More Than 600 Volts.

(1) General. Branch-circuit conductors shall have an ampacity not less than the maximum load to be served. Where a branch circuit supplies continuous loads or any combination of continuous and noncontinuous loads, the minimum branch-circuit conductor size, before the application of any adjustment or correction factors, shall have an allowable ampacity not less than the noncontinuous load plus 125 percent of the continuous load.

(2) Branch Circuits with More than One Receptacle. Conductors of branch circuits supplying more than one receptacle for cord-and-plug-connected portable loads shall have an ampacity of not less than the rating of the branch circuit.

(3) Household Ranges and Cooking Appliances. Branch-circuit conductors supplying household ranges, wall-mounted ovens, counter-mounted cooking units, and other household cooking appliances shall have an ampacity not less than the rating of the branch circuit and not less than the maximum load to be served. For ranges of 8-3/4 kW or more rating, the minimum branch-circuit rating shall be 40 amperes.

Exception No. 1: Conductors tapped from a 50-ampere branch circuit supplying electric ranges, wall-mounted electric ovens, and counter-mounted electric cooking units shall have an ampacity of not less than 20 amperes and shall be sufficient for the load to be served. These tap conductors include any conductors that are a part of the leads supplied with the appliance that are smaller than the branch-circuit conductors. The taps shall not be longer than necessary for servicing the appliance.

Exception No. 2: The neutral conductor of a 3-wire branch circuit supplying a household electric range, a wall-mounted oven, or a counter-mounted cooking unit shall be permitted to be smaller than the ungrounded conductors where the maximum demand of a range of 8-3/4-kW or more rating has been calculated according to Column C of Table 220.55, but such conductor shall have an ampacity of not less than 70 percent of the branch-circuit rating and shall not be smaller than 10 AWG.

(4) Other Loads. Branch-circuit conductors that supply loads other than those specified in 210.2 and other than cooking appliances as covered in 210.19(A)(3) shall have an ampacity sufficient for the loads served and shall not be smaller than 14 AWG.

210.20 Overcurrent Protection. Branch-circuit conductors and equipment shall be protected by overcurrent protective devices that have a rating or setting that complies with 210.20(A) through (D).

(A) Continuous and Noncontinuous Loads. Where a branch circuit supplies continuous loads or any combination of continuous and noncontinuous loads, the rating of the overcurrent device shall not be less than the noncontinuous load plus 125 percent of the continuous load.

(B) Conductor Protection. Conductors shall be protected in accordance with 240.4.

(D) Outlet Devices. The rating or setting shall not exceed that specified in 210.21 for outlet devices.

210.21 Outlet Devices. Outlet devices shall have an ampere rating that is not less than the load to be served and shall comply with 210.21(A) and (B).

(B) Receptacles.

(1) Single Receptacle on an Individual Branch Circuit. A single receptacle installed on an individual branch circuit shall have an ampere rating not less than that of the branch circuit.

(2) Total Cord-and-Plug-Connected Load. Where connected to a branch circuit supplying two or more receptacles or outlets, a receptacle shall not supply a total cord-and-plug-connected load in excess of the maximum specified in Table 210.21(B)(2).

(3) Receptacle Ratings. Where connected to a branch circuit supplying two or more receptacles or outlets, receptacle ratings shall conform to the values listed in Table 210.21(B)(3), or, where rated higher than 50 amperes, the receptacle rating shall not be less than the branch-circuit rating.

Table 210.21(B)(2) Maximum Cord-and-Plug-Connected Load to Receptacle

Circuit Rating (Amperes)	Receptacle Rating (Amperes)	Maximum Load (Amperes)
15 or 20	15	12
20	20	16
30	30	24

**Table 210.21(B)(3) Receptacle Ratings
for Various Size Circuits**

Circuit Rating (Amperes)	Receptacle Rating (Amperes)
15	Not over 15
20	15 or 20
30	30
40	40 or 50
50	50

(4) Range Receptacle Rating. The ampere rating of a range receptacle shall be permitted to be based on a single range demand load as specified in Table 220.55.

210.23 Permissible Loads. In no case shall the load exceed the branch-circuit ampere rating. An individual branch circuit shall be permitted to supply any load for which it is rated. A branch circuit supplying two or more outlets or receptacles shall supply only the loads specified according to its size as specified in 210.23(A) through (D) and as summarized in 210.24 and Table 210.24.

(A) 15- and 20-Ampere Branch Circuits. A 15- or 20-ampere branch circuit shall be permitted to supply lighting units or other utilization equipment, or a combination of both, and shall comply with 210.23(A)(1) and (A)(2).

Exception: The small-appliance branch circuits, laundry branch circuits, and bathroom branch circuits required in a dwelling unit(s) by 210.11(C)(1), (C)(2), and (C)(3) shall supply only the receptacle outlets specified in that section.

(1) Cord-and-Plug-Connected Equipment Not Fastened in Place. The rating of any one cord-and-plug-connected utilization equipment not fastened in place shall not exceed 80 percent of the branch-circuit ampere rating.

(2) Utilization Equipment Fastened in Place. The total rating of utilization equipment fastened in place, other than luminaires, shall not exceed 50 percent of the branch-circuit ampere rating where lighting units, cord-and-plug-connected utilization equipment not fastened in place, or both, are also supplied.

(B) 30-Ampere Branch Circuits. A 30-ampere branch circuit shall be permitted to supply fixed lighting units with heavy-duty lampholders in other than a dwelling unit(s) or utilization equipment in any occupancy. A rating of any one cord-and-plug-connected utilization equipment shall not exceed 80 percent of the branch-circuit ampere rating.

(C) 40- and 50-Ampere Branch Circuits. A 40- or 50-ampere branch circuit shall be permitted to supply cooking appliances that are fastened in place in any occupancy.

(D) Branch Circuits Larger Than 50 Amperes. Branch circuits larger than 50 amperes shall supply only nonlighting outlet loads.

210.24 Branch-Circuit Requirements—Summary. The requirements for circuits that have two or more outlets or receptacles, other than the receptacle circuits of 210.11(C)(1), (C)(2), and (C)(3), are summarized in Table 210.24. This table provides only a summary of minimum requirements. See 210.19, 210.20, and 210.21 for the specific requirements applying to branch circuits.

210.25 Branch Circuits in Buildings with More Than One Occupancy.

(A) Dwelling Unit Branch Circuits. Branch circuits in each dwelling unit shall supply only loads within that dwelling unit or loads associated only with that dwelling unit.

(B) Common Area Branch Circuits. Branch circuits installed for the purpose of lighting, central alarm, signal, communications, or other purposes for public or common areas of a two-family dwelling, a multifamily dwelling, or a multi-occupancy building shall not be supplied from equipment that supplies an individual dwelling unit or tenant space.

III. REQUIRED OUTLETS

(See Chapter 17 for required receptacles and Chapter 19 for required lighting outlets.)

Table 210.24 Summary of Branch-Circuit Requirements

Circuit Rating	15 A	20 A	30 A	40 A	50 A
Conductors (min. size):					
Circuit wires[1]	14	12	10	8	6
Taps	14	14	14	12	12
Fixture wires and cords—see 240.5					
Overcurrent Protection	**15 A**	**20 A**	**30 A**	**40 A**	**50 A**
Outlet devices:					
Lampholders permitted	Any type	Any type	Heavy duty	Heavy duty	Heavy duty
Receptacle rating[2]	15 max. A	15 or 20 A	30 A	40 or 50 A	50 A
Maximum Load	**15 A**	**20 A**	**30 A**	**40 A**	**50 A**
Permissible load	See 210.23(A)	See 210.23(A)	See 210.23(B)	See 210.23(C)	See 210.23(C)
		See 210.23(B)	See 210.23(C)		

[1]These gauges are for copper conductors.

[2]For receptacle rating of cord-connected electric-discharge luminaires, see 410.62(C).

CHAPTER 17

RECEPTACLE PROVISIONS

INTRODUCTION

This chapter contains provisions from Article 210—Branch Circuits and Article 406—Receptacles, Cord Connectors, and Attachment Plugs (Caps). The provisions from Article 210 covered in this chapter include 210.50 through 210.63. In those sections are specifications for installing receptacle outlets for both general and specific locations. General locations include: kitchens, family rooms, dining rooms, living rooms, bedrooms, and similar rooms and areas. Specific locations include: bathrooms, hallways, basements and garages, laundry areas, and kitchen counter-top areas.

Article 406 covers the rating, type, and installation of receptacles, cord connectors, and attachment plugs (cord caps). Article 406 also contains requirements on replacing receptacles installed in existing dwelling units.

ARTICLE 210
Branch Circuits

III. REQUIRED OUTLETS

210.50 General. Receptacle outlets shall be installed as specified in 210.52 through 210.63.

(A) Cord Pendants. A cord connector that is supplied by a permanently connected cord pendant shall be considered a receptacle outlet.

(B) Cord Connections. A receptacle outlet shall be installed wherever flexible cords with attachment plugs are used. Where flexible cords are permitted to be permanently connected, receptacles shall be permitted to be omitted for such cords.

(C) Appliance Receptacle Outlets. Appliance receptacle outlets installed in a dwelling unit for specific appliances, such as laundry equipment, shall be installed within 1.8 m (6 ft) of the intended location of the appliance.

210.52 Dwelling Unit Receptacle Outlets. This section provides requirements for 125-volt, 15- and 20-ampere receptacle outlets. The receptacles required by this section shall be in addition to any receptacle that is:

(1) Part of a luminaire or appliance, or
(2) Controlled by a wall switch in accordance with 210.70(A)(1), Exception No. 1, or
(3) Located within cabinets or cupboards, or
(4) Located more than 1.7 m (5-1/2 ft) above the floor

Permanently installed electric baseboard heaters equipped with factory-installed receptacle outlets or outlets provided as a separate assembly by the manufacturer shall be permitted as the required outlet or outlets for the wall space utilized by such permanently installed heaters. Such receptacle outlets shall not be connected to the heater circuits.

Informational Note: Listed baseboard heaters include instructions that may not permit their installation below receptacle outlets.

(A) General Provisions. In every kitchen, family room, dining room, living room, parlor, library, den, sunroom, bedroom, recreation room, or similar room or area of dwelling units, receptacle outlets shall be installed in accordance with the general provisions specified in 210.52(A)(1) through (A)(3).

(1) Spacing. Receptacles shall be installed such that no point measured horizontally along the floor line of any wall space is more than 1.8 m (6 ft) from a receptacle outlet.

(2) Wall Space. As used in this section, a wall space shall include the following:

(1) Any space 600 mm (2 ft) or more in width (including space measured around corners) and unbroken along the floor line by doorways and similar openings, fireplaces, and fixed cabinets

(2) The space occupied by fixed panels in exterior walls, excluding sliding panels

(3) The space afforded by fixed room dividers, such as freestanding bar-type counters or railings

(3) Floor Receptacles. Receptacle outlets in floors shall not be counted as part of the required number of receptacle outlets unless located within 450 mm (18 in.) of the wall.

(4) Countertop Receptacles. Receptacles installed for countertop surfaces as specified in 210.52(C) shall not be considered as the receptacles required by 210.52(A).

(B) Small Appliances.

(1) Receptacle Outlets Served. In the kitchen, pantry, breakfast room, dining room, or similar area of a dwelling unit, the two or more 20-ampere small-appliance branch circuits required by 210.11(C)(1) shall serve all wall and floor receptacle outlets covered by 210.52(A), all countertop outlets covered by 210.52(C), and receptacle outlets for refrigeration equipment.

Exception No. 1: In addition to the required receptacles specified by 210.52, switched receptacles supplied from a general-purpose branch circuit as defined in 210.70(A)(1), Exception No. 1, shall be permitted.

Exception No. 2: The receptacle outlet for refrigeration equipment shall be permitted to be supplied from an individual branch circuit rated 15 amperes or greater.

(2) No Other Outlets. The two or more small-appliance branch circuits specified in 210.52(B)(1) shall have no other outlets.

Exception No. 1: A receptacle installed solely for the electrical supply to and support of an electric clock in any of the rooms specified in 210.52(B)(1).

Exception No. 2: Receptacles installed to provide power for supplemental equipment and lighting on gas-fired ranges, ovens, or counter-mounted cooking units.

(3) Kitchen Receptacle Requirements. Receptacles installed in a kitchen to serve countertop surfaces shall be supplied by not fewer than two small-appliance branch circuits, either or both of which shall also be permitted to supply receptacle outlets in the same kitchen and in other rooms specified in 210.52(B)(1). Additional small-appliance branch circuits shall be permitted to supply receptacle outlets in the kitchen and other rooms specified in 210.52(B)(1). No small-appliance branch circuit shall serve more than one kitchen.

(C) Countertops. In kitchens, pantries, breakfast rooms, dining rooms, and similar areas of dwelling units, receptacle outlets for countertop spaces shall be installed in accordance with 210.52(C)(1) through (C)(5).

(1) Wall Countertop Spaces. A receptacle outlet shall be installed at each wall countertop space that is 300 mm (12 in.) or wider. Receptacle outlets shall be installed so that no point along the wall line is more than 600 mm (24 in.) measured horizontally from a receptacle outlet in that space.

Exception: Receptacle outlets shall not be required on a wall directly behind a range, counter-mounted cooking unit, or sink in the installation described in Figure 210.52(C)(1).

(2) Island Countertop Spaces. At least one receptacle shall be installed at each island countertop space with a long dimension of 600 mm (24 in.) or greater and a short dimension of 300 mm (12 in.) or greater.

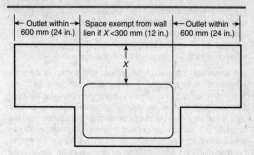

Range, counter-mounted cooking unit extending from face of counter

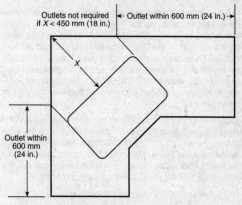

Range, counter-mounted cooking unit mounted in corner

Figure 210.52(C)(1) Determination of Area Behind a Range, or Counter-Mounted Cooking Unit or Sink.

(3) Peninsular Countertop Spaces. At least one receptacle outlet shall be installed at each peninsular countertop space with a long dimension of 600 mm (24 in.) or greater and a short dimension of 300 mm (12 in.) or greater. A peninsular countertop is measured from the connecting edge.

(4) Separate Spaces. Countertop spaces separated by rangetops, refrigerators, or sinks shall be considered as separate countertop spaces in applying the requirements of 210.52(C)(1). If a range, counter-mounted cooking unit, or sink is installed in an island or peninsular countertop and the depth of the countertop behind the range, counter-mounted cooking unit, or sink is less than 300 mm (12 in.), the range, counter-mounted cooking unit, or sink shall be considered to divide the countertop space into two separate countertop spaces. Each separate countertop space shall comply with the applicable requirements in 210.52(C).

(5) Receptacle Outlet Location. Receptacle outlets shall be located on or above, but not more than 500 mm (20 in.) above, the countertop. Receptacle outlet assemblies listed for the application shall be permitted to be installed in countertops. Receptacle outlets rendered not readily accessible by appliances fastened in place, appliance garages, sinks, or rangetops as covered in 210.52(C)(1), Exception, or appliances occupying dedicated space shall not be considered as these required outlets.

Exception to (5): To comply with the conditions specified in (1) or (2), receptacle outlets shall be permitted to be mounted not more than 300 mm (12 in.) below the countertop. Receptacles mounted below a countertop in accordance with this exception shall not be located where the countertop extends more than 150 mm (6 in.) beyond its support base.

(1) Construction for the physically impaired

(2) On island and peninsular countertops where the countertop is flat across its entire surface (no backsplashes, dividers, etc.) and there are no means to mount a receptacle within 500 mm (20 in.) above the countertop, such as an overhead cabinet

(D) Bathrooms. In dwelling units, at least one receptacle outlet shall be installed in bathrooms within 900 mm (3 ft) of the outside edge of each basin. The receptacle outlet shall be located on a wall or partition that is adjacent to the basin or basin countertop, located on the countertop, or installed on the side or face of the basin cabinet not more than 300 mm (12 in.) below the countertop. Receptacle outlet assemblies listed for the application shall be permitted to be installed in the countertop.

(E) Outdoor Outlets. Outdoor receptacle outlets shall be installed in accordance with (E)(1) through (E)(3). *[See 210.8(A)(3).]*

(1) One-Family and Two-Family Dwellings. For a one-family dwelling and each unit of a two-family dwelling that is at grade level, at least one receptacle outlet accessible while standing at grade level and located not more than 2.0 m (6-1/2 ft) above grade shall be installed at the front and back of the dwelling.

(2) Multifamily Dwellings. For each dwelling unit of a multifamily dwelling where the dwelling unit is located at grade level and provided with individual exterior entrance/egress, at least one receptacle outlet accessible from grade level and not more than 2.0 m (6-1/2 ft) above grade shall be installed.

(3) Balconies, Decks, and Porches. Balconies, decks, and porches that are accessible from inside the dwelling unit shall have at least one receptacle outlet installed within the perimeter of the balcony, deck, or porch. The receptacle shall not be located more than 2.0 m (6-1/2 ft) above the balcony, deck, or porch surface.

(F) Laundry Areas. In dwelling units, at least one receptacle outlet shall be installed for the laundry.

Exception No. 1: In a dwelling unit that is an apartment or living area in a multifamily building where laundry facilities are provided on the premises and are available to all building occupants, a laundry receptacle shall not be required.

Exception No. 2: In other than one-family dwellings where laundry facilities are not to be installed or permitted, a laundry receptacle shall not be required.

(G) Basements, Garages, and Accessory Buildings. For a one-family dwelling, the following provisions shall apply:

(1) At least one receptacle outlet, in addition to those for specific equipment, shall be installed in each basement, in each attached garage, and in each detached garage or accessory building with electric power.

(2) Where a portion of the basement is finished into one or more habitable rooms, each separate unfinished portion shall have a receptacle outlet installed in accordance with this section.

(H) Hallways. In dwelling units, hallways of 3.0 m (10 ft) or more in length shall have at least one receptacle outlet.

As used in this subsection, the hallway length shall be considered the length along the centerline of the hallway without passing through a doorway.

(I) Foyers. Foyers that are not part of a hallway in accordance with 210.52(H) and that have an area that is greater than 5.6 m² (60 ft²) shall have a receptacle(s) located in each wall space 900 mm (3 ft) or more in width and unbroken by doorways, floor-to-ceiling windows, and similar openings.

210.63 Heating, Air-Conditioning, and Refrigeration Equipment Outlet. A 125-volt, single-phase, 15- or 20-ampere-rated receptacle outlet shall be installed at an accessible location for the servicing of heating, air-conditioning, and refrigeration equipment. The receptacle shall be located on the same level and within 7.5 m (25 ft) of the heating, air-conditioning, and refrigeration equipment. The receptacle outlet shall not be connected to the load side of the equipment disconnecting means.

Exception: A receptacle outlet shall not be required at one- and two-family dwellings for the service of evaporative coolers.

ARTICLE 406

Receptacles, Cord Connectors, and Attachment Plugs (Caps)

406.1 Scope. This article covers the rating, type, and installation of receptacles, cord connectors, and attachment plugs (cord caps).

406.3 Receptacle Rating and Type.

(A) Receptacles. Receptacles shall be listed and marked with the manufacturer's name or identification and voltage and ampere ratings.

(B) Rating. Receptacles and cord connectors shall be rated not less than 15 amperes, 125 volts, or 15 amperes, 250 volts, and shall be of a type not suitable for use as lampholders.

(C) Receptacles for Aluminum Conductors. Receptacles rated 20 amperes or less and designed for the direct connection of aluminum conductors shall be marked CO/ALR.

(D) Isolated Ground Receptacles. Receptacles incorporating an isolated grounding conductor connection intended for the reduction of electrical noise (electromagnetic interference) as permitted in 250.146(D) shall be identified by an orange triangle located on the face of the receptacle.

(1) Isolated Equipment Grounding Conductor Required. Receptacles so identified shall be used only with equipment grounding conductors that are isolated in accordance with 250.146(D).

(2) Installation in Nonmetallic Boxes. Isolated ground receptacles installed in nonmetallic boxes shall be covered with a nonmetallic faceplate.

Exception: Where an isolated ground receptacle is installed in a nonmetallic box, a metal faceplate shall be permitted if the box contains a feature or accessory that permits the effective grounding of the faceplate.

406.4 General Installation Requirements. Receptacle outlets shall be located in branch circuits in accordance with Part III of Article 210. General installation requirements shall be in accordance with 406.4(A) through (F).

(A) Grounding Type. Receptacles installed on 15- and 20-ampere branch circuits shall be of the grounding type. Grounding-type receptacles shall be installed only on circuits of the voltage class and current for which they are rated, except as provided in Table 210.21(B)(2) and Table 210.21(B)(3).

Exception: Nongrounding-type receptacles installed in accordance with 406.4(D).

(B) To Be Grounded. Receptacles and cord connectors that have equipment grounding conductor contacts shall have those contacts connected to an equipment grounding conductor.

Exception No. 2: Replacement receptacles as permitted by 406.4(D).

(C) Methods of Grounding. The equipment grounding conductor contacts of receptacles and cord connectors shall be grounded by connection to the equipment grounding conductor of the circuit supplying the receptacle or cord connector.

The branch-circuit wiring method shall include or provide an equipment grounding conductor to which the equipment grounding conductor contacts of the receptacle or cord connector are connected.

(D) Replacements. Replacement of receptacles shall comply with 406.4(D)(1) through (D)(6), as applicable.

(1) Grounding-Type Receptacles. Where a grounding means exists in the receptacle enclosure or an equipment grounding conductor is installed in accordance with 250.130(C), grounding-type receptacles shall be used and shall be connected to the equipment grounding conductor in accordance with 406.4(C) or 250.130(C).

(2) Non–Grounding-Type Receptacles. Where attachment to an equipment grounding conductor does not exist in the receptacle enclosure, the installation shall comply with (D)(2)(a), (D)(2)(b), or (D)(2)(c).

(a) A non–grounding-type receptacle(s) shall be permitted to be replaced with another non–grounding-type receptacle(s).

(b) A non–grounding-type receptacle(s) shall be permitted to be replaced with a ground-fault circuit interrupter-type of receptacle(s). These receptacles shall be marked "No Equipment Ground." An equipment grounding conductor shall not be connected from the ground-fault circuit-interrupter-type receptacle to any outlet supplied from the ground-fault circuit-interrupter receptacle.

(c) A non–grounding-type receptacle(s) shall be permitted to be replaced with a grounding-type receptacle(s) where supplied through a ground-fault circuit interrupter. Grounding-type receptacles supplied through the ground-fault circuit interrupter shall be marked "GFCI Protected" and "No Equipment Ground." An equipment grounding conductor shall not be connected between the grounding-type receptacles.

(3) Ground-Fault Circuit-Interrupters. Ground-fault circuit-interrupter protected receptacles shall be provided where replacements are made at receptacle outlets that are required to be so protected elsewhere in this *Code*.

(4) Arc-Fault Circuit-Interrupter Protection. Where a receptacle outlet is supplied by a branch circuit that requires arc-fault circuit interrupter protection as specified elsewhere in this *Code*, a replacement receptacle at this outlet shall be one of the following:

(1) A listed outlet branch circuit type arc-fault circuit interrupter receptacle
(2) A receptacle protected by a listed outlet branch circuit type arc-fault circuit interrupter type receptacle
(3) A receptacle protected by a listed combination type arc-fault circuit interrupter type circuit breaker
    This requirement becomes effective January 1, 2014.

(5) Tamper-Resistant Receptacles. Listed tamper-resistant receptacles shall be provided where replacements are made at receptacle outlets that are required to be tamper-resistant elsewhere in this *Code*.

(6) Weather-Resistant Receptacles. Weather-resistant receptacles shall be provided where replacements are made at receptacle outlets that are required to be so protected elsewhere in this *Code*.

406.5 Receptacle Mounting. Receptacles shall be mounted in boxes or assemblies designed for the purpose, and such boxes or assemblies shall be securely fastened in place unless otherwise permitted elsewhere in this *Code*.

(A) Boxes That Are Set Back. Receptacles mounted in boxes that are set back from the finished surface as permitted in 314.20 shall be installed such that the mounting yoke or strap of the receptacle is held rigidly at the finished surface.

(B) Boxes That Are Flush. Receptacles mounted in boxes that are flush with the finished surface or project therefrom shall be installed such that the mounting yoke or strap of the receptacle is held rigidly against the box or box cover.

(C) Receptacles Mounted on Covers. Receptacles mounted to and supported by a cover shall be held rigidly against the cover by more than one screw or shall be a device assembly or box cover listed and identified for securing by a single screw.

(D) Position of Receptacle Faces. After installation, receptacle faces shall be flush with or project from faceplates of insulating material and shall project a minimum of 0.4 mm (0.015 in.) from metal faceplates.

Exception: Listed kits or assemblies encompassing receptacles and nonmetallic faceplates that cover the receptacle face, where the plate cannot be installed on any other receptacle, shall be permitted.

(E) Receptacles in Countertops and Similar Work Surfaces in Dwelling Units. Receptacles shall not be installed in a face-up position in countertops or similar work surfaces.

(F) Exposed Terminals. Receptacles shall be enclosed so that live wiring terminals are not exposed to contact.

406.6 Receptacle Faceplates (Cover Plates). Receptacle faceplates shall be installed so as to completely cover the opening and seat against the mounting surface.

Receptacle faceplates mounted inside a box having a recess-mounted receptacle shall effectively close the opening and seat against the mounting surface.

(A) Thickness of Metal Faceplates. Metal faceplates shall be of ferrous metal not less than 0.76 mm (0.030 in.) in thickness or of nonferrous metal not less than 1.02 mm (0.040 in.) in thickness.

(B) Grounding. Metal faceplates shall be grounded.

(C) Faceplates of Insulating Material. Faceplates of insulating material shall be noncombustible and not less than 2.54 mm (0.10 in.) in thickness but shall be permitted to be less than 2.54 mm (0.10 in.) in thickness if formed or reinforced to provide adequate mechanical strength.

406.7 Attachment Plugs, Cord Connectors, and Flanged Surface Devices. All attachment plugs, cord connectors, and flanged surface devices (inlets and outlets) shall be listed and marked with the manufacturer's name or identification and voltage and ampere ratings.

(A) Construction of Attachment Plugs and Cord Connectors. Attachment plugs and cord connectors shall be constructed so that there are no exposed current-carrying parts except the prongs, blades, or pins. The cover for wire terminations shall be a part that is essential for the operation of an attachment plug or connector (dead-front construction).

(B) Connection of Attachment Plugs. Attachment plugs shall be installed so that their prongs, blades, or pins are not energized unless inserted into an energized receptacle or cord connectors. No receptacle shall be installed so as to require the insertion of an energized attachment plug as its source of supply.

406.9 Receptacles in Damp or Wet Locations.

(A) Damp Locations. A receptacle installed outdoors in a location protected from the weather or in other damp locations shall have an enclosure for the receptacle that is weatherproof when the receptacle is covered (attachment plug cap not inserted and receptacle covers closed).

An installation suitable for wet locations shall also be considered suitable for damp locations.

A receptacle shall be considered to be in a location protected from the weather where located under roofed open porches, canopies, marquees, and the like, and will not be subjected to a beating rain or water runoff. All 15- and 20-ampere, 125- and 250-volt nonlocking receptacles shall be a listed weather-resistant type.

(B) Wet Locations.

(1) 15- and 20-Ampere Receptacles in a Wet Location. 15- and 20-ampere, 125- and 250-volt receptacles installed in a wet location shall have an enclosure that is weatherproof whether or not the attachment plug cap is inserted. For other than one- or two-family dwellings, an outlet box hood installed for this purpose shall be listed, and where installed on an enclosure supported from grade as described in 314.23(B) or as described in 314.23(F) shall be identified as "extra-duty." All 15- and 20-ampere, 125- and 250-volt nonlocking-type receptacles shall be listed weather-resistant type.

(2) Other Receptacles. All other receptacles installed in a wet location shall comply with (B)(2)(a) or (B)(2)(b).

(a) A receptacle installed in a wet location, where the product intended to be plugged into it is not attended while in use, shall have an enclosure that is weatherproof with the attachment plug cap inserted or removed.

(b) A receptacle installed in a wet location where the product intended to be plugged into it will be attended while in use (e.g., portable tools) shall have an enclosure that is weatherproof when the attachment plug is removed.

(C) Bathtub and Shower Space. Receptacles shall not be installed within or directly over a bathtub or shower stall.

(E) Flush Mounting with Faceplate. The enclosure for a receptacle installed in an outlet box flush-mounted in a finished surface shall be made weatherproof by means of a weatherproof faceplate assembly that provides a watertight connection between the plate and the finished surface.

406.11 Connecting Receptacle Grounding Terminal to Box. The connection of the receptacle grounding terminal shall comply with 250.146.

406.12 Tamper-Resistant Receptacles in Dwelling Units. In all areas specified in 210.52, all nonlocking-type 125-volt, 15- and 20-ampere receptacles shall be listed tamper-resistant receptacles.

Exception: Receptacles in the following locations shall not be required to be tamper-resistant:

(1) *Receptacles located more than 1.7 m (5-1/2 ft) above the floor.*
(2) *Receptacles that are part of a luminaire or appliance.*
(3) *A single receptacle or a duplex receptacle for two appliances located within dedicated space for each appliance that, in normal use, is not easily moved from one place to another and that is cord-and-plug connected in accordance with 400.7(A)(6), (A)(7), or (A)(8).*
(4) *Nongrounding receptacles used for replacements as permitted in 406.4(D)(2)(a).*

CHAPTER 18

SWITCHES

INTRODUCTION

This chapter contains provisions from Article 404—Switches. Unlike the disconnect switches covered in Chapter 8, this chapter covers general-use snap switches. These switches are generally installed in device boxes or on box covers. Provisions from this article cover general-use snap switches (single-pole, three-way, four-way, double-pole, etc.), dimmers, and similar control switches. Specifications pertaining to the required locations for switches are in Article 210 (see Chapter 19).

ARTICLE 404
Switches

I. INSTALLATION

404.1 Scope. The provisions of this article apply to all switches, switching devices, and circuit breakers used as switches, operating at 600 volts and below, unless specifically referenced elsewhere in this *Code* for higher voltages.

404.2 Switch Connections.

(A) Three-Way and Four-Way Switches. Three-way and four-way switches shall be wired so that all switching is done only in the ungrounded circuit conductor. Where in metal raceways or metal-armored cables, wiring between switches and outlets shall be in accordance with 300.20(A).

Exception: Switch loops shall not require a grounded conductor.

(B) Grounded Conductors. Switches or circuit breakers shall not disconnect the grounded conductor of a circuit.

(C) Switches Controlling Lighting Loads. Where switches control lighting loads supplied by a grounded general purpose branch circuit, the grounded circuit conductor for the controlled lighting circuit shall be provided at the switch location.

Exception: The grounded circuit conductor shall be permitted to be omitted from the switch enclosure where either of the following conditions in (1) or (2) apply:

(1) Conductors for switches controlling lighting loads enter the box through a raceway. The raceway shall have sufficient cross-sectional area to accommodate the extension of the grounded circuit conductor of the lighting circuit to the switch location whether or not the conductors in the raceway are required to be increased in size to comply with 310.15(B)(3)(a).

(2) Cable assemblies for switches controlling lighting loads enter the box through a framing cavity that is open at the top or bottom on the same floor level, or through a wall, floor, or ceiling that is unfinished on one side.

404.4 Damp or Wet Locations.

(A) Surface-Mounted Switch or Circuit Breaker. A surface-mounted switch or circuit breaker shall be enclosed in a weatherproof enclosure or cabinet that shall comply with 312.2.

Note: See 312.2 following 408.37 in Chapter 8.

(B) Flush-Mounted Switch or Circuit Breaker. A flush-mounted switch or circuit breaker shall be equipped with a weatherproof cover.

(C) Switches in Tub or Shower Spaces. Switches shall not be installed within tubs or shower spaces unless installed as part of a listed tub or shower assembly.

404.8 Accessibility and Grouping.

(A) Location. All switches and circuit breakers used as switches shall be located so that they may be operated from a readily accessible place. They shall be installed such that the center of the grip of the operating handle of the switch or circuit breaker, when in its highest position, is not more than 2.0 m (6 ft 7 in.) above the floor or working platform.

(C) Multipole Snap Switches. A multipole, general-use snap switch shall not be permitted to be fed from more than a single circuit unless it is listed and marked as a two-circuit or three-circuit switch, or unless its voltage rating is not less than the nominal line-to-line voltage of the system supplying the circuits.

404.9 Provisions for General-Use Snap Switches.

(A) Faceplates. Faceplates provided for snap switches mounted in boxes and other enclosures shall be installed so as to completely cover the opening and, where the switch is flush mounted, seat against the finished surface.

(B) Grounding. Snap switches, including dimmer and similar control switches, shall be connected to an equipment grounding conductor and shall provide a means to connect metal faceplates to the equipment grounding conductor, whether or not a metal faceplate is installed. Snap switches shall be considered to be part of an effective ground-fault current path if either of the following conditions is met:

(1) The switch is mounted with metal screws to a metal box or metal cover that is connected to an equipment grounding conductor or to a nonmetallic box with integral means for connecting to an equipment grounding conductor.

(2) An equipment grounding conductor or equipment bonding jumper is connected to an equipment grounding termination of the snap switch.

Exception No. 1 to (B): Where no means exists within the snap-switch enclosure for connecting to the equipment grounding conductor, or where the wiring method does not include or provide an equipment grounding conductor, a snap switch without a connection to an equipment grounding conductor shall be permitted for replacement purposes only. A snap switch wired under the provisions of this exception and located within 2.5 m (8 ft) vertically, or 1.5 m (5 ft) horizontally, of ground or exposed grounded metal objects shall be provided with a faceplate of nonconducting noncombustible material with nonmetallic attachment screws, unless the switch mounting strap or yoke is nonmetallic or the circuit is protected by a ground-fault circuit interrupter.

Exception No. 2 to (B): Listed kits or listed assemblies shall not be required to be connected to an equipment grounding conductor if all of the following conditions are met:

(1) The device is provided with a nonmetallic faceplate that cannot be installed on any other type of device,

(2) The device does not have mounting means to accept other configurations of faceplates,

(3) The device is equipped with a nonmetallic yoke, and

(4) All parts of the device that are accessible after installation of the faceplate are manufactured of nonmetallic materials.

Exception No. 3 to (B): A snap switch with integral nonmetallic enclosure complying with 300.15(E) shall be permitted without a connection to an equipment grounding conductor.

404.10 Mounting of Snap Switches.

(A) Surface Type. Snap switches used with open wiring on insulators shall be mounted on insulating material that separates the conductors at least 13 mm (1/2 in.) from the surface wired over.

(B) Box Mounted. Flush-type snap switches mounted in boxes that are set back of the finished surface as permitted in 314.20 shall be installed so that the extension plaster ears are seated against the surface. Flush-type snap switches mounted in boxes that are flush with the finished surface or project from it shall be installed so that the mounting yoke or strap of the switch is seated against the box.

404.14 Rating and Use of Snap Switches. Snap switches shall be used within their ratings and as indicated in 404.14(A) through (F).

(A) Alternating-Current General-Use Snap Switch. A form of general-use snap switch suitable only for use on ac circuits for controlling the following:

(1) Resistive and inductive loads not exceeding the ampere rating of the switch at the voltage involved

(2) Tungsten-filament lamp loads not exceeding the ampere rating of the switch at 120 volts

(3) Motor loads not exceeding 80 percent of the ampere rating of the switch at its rated voltage

(C) CO/ALR Snap Switches. Snap switches rated 20 amperes or less directly connected to aluminum conductors shall be listed and marked CO/ALR.

(E) Dimmer Switches. General-use dimmer switches shall be used only to control permanently installed incandescent luminaires unless listed for the control of other loads and installed accordingly.

(F) Cord-and-Plug-Connected Loads. Where a snap switch is used to control cord-and-plug-connected equipment on a general-purpose branch circuit, each snap switch controlling receptacle outlets or cord connectors that are supplied by permanently connected cord pendants shall be rated at not less than the rating of the maximum permitted ampere rating or setting of the overcurrent device protecting the receptacles or cord connectors, as provided in 210.21(B).

Exception: Where a snap switch is used to control not more than one receptacle on a branch circuit, the switch shall be permitted to be rated at not less than the rating of the receptacle.

CHAPTER 19

LIGHTING REQUIREMENTS

INTRODUCTION

This chapter contains provisions from Article 210—Branch Circuits; Article 410—Luminaires, Lampholders, and Lamps; and Article 411—Lighting Systems Operating at 30 Volts or Less. This chapter covers the last section of Article 210, which provides requirements for lighting outlets and switches for dwelling units.

Article 410 covers luminaires, lampholders, pendants, incandescent filament lamps, and wiring and equipment forming part of such lamps, luminaires, and lighting installations. Contained in this pocket guide are eight of the fifteen parts that are included in Article 410. The eight parts are: I General; II Luminaire Locations; III Provisions at Luminaire Outlet Boxes, Canopies, and Pans; IV Luminaire Supports; V Grounding; VI Wiring of Luminaires; X Special Provisions for Flush and Recessed Luminaires; and XIV Lighting Track.

Article 411 covers lighting systems operating at 30 volts or less and their associated components.

ARTICLE 210
Branch Circuits

III. REQUIRED OUTLETS

210.70 Lighting Outlets Required. Lighting outlets shall be installed where specified in 210.70(A), (B), and (C).

(A) Dwelling Units. In dwelling units, lighting outlets shall be installed in accordance with 210.70(A)(1), (A)(2), and (A)(3).

(1) Habitable Rooms. At least one wall switch–controlled lighting outlet shall be installed in every habitable room and bathroom.

Exception No. 1: In other than kitchens and bathrooms, one or more receptacles controlled by a wall switch shall be permitted in lieu of lighting outlets.

Exception No. 2: Lighting outlets shall be permitted to be controlled by occupancy sensors that are (1) in addition to wall switches or (2) located at a customary wall switch location and equipped with a manual override that will allow the sensor to function as a wall switch.

(2) Additional Locations. Additional lighting outlets shall be installed in accordance with (A)(2)(a), (A)(2)(b), and (A)(2)(c).

(a) At least one wall switch–controlled lighting outlet shall be installed in hallways, stairways, attached garages, and detached garages with electric power.

(b) For dwelling units, attached garages, and detached garages with electric power, at least one wall switch–controlled lighting outlet shall be installed to provide illumination on the exterior side of outdoor entrances or exits with grade level access. A vehicle door in a garage shall not be considered as an outdoor entrance or exit.

(c) Where one or more lighting outlet(s) are installed for interior stairways, there shall be a wall switch at each floor level, and landing level that includes an entryway, to control the lighting outlet(s) where the stairway between floor levels has six risers or more.

Exception to (A)(2)(a), (A)(2)(b), and (A)(2)(c): In hallways, in stairways, and at outdoor entrances, remote, central, or automatic control of lighting shall be permitted.

(3) Storage or Equipment Spaces. For attics, underfloor spaces, utility rooms, and basements, at least one lighting outlet containing a switch or controlled by a wall switch shall be installed where these spaces are used for storage or contain equipment requiring servicing. At least one point of control shall be at the usual point of entry to these spaces. The lighting outlet shall be provided at or near the equipment requiring servicing.

ARTICLE 410

Luminaires, Lampholders, and Lamps

I. GENERAL

410.1 Scope. This article covers luminaires, portable luminaires, lampholders, pendants, incandescent filament lamps, arc lamps, electric-discharge lamps, decorative lighting products, lighting accessories for temporary seasonal and holiday use, portable flexible lighting products, and the wiring and equipment forming part of such products and lighting installations.

410.2 Definitions.

Closet Storage Space. The volume bounded by the sides and back closet walls and planes extending from the closet floor vertically to a height of 1.8 m (6 ft) or to the highest clothes-hanging rod and parallel to the walls at a horizontal distance of 600 mm (24 in.) from the sides and back of the closet walls, respectively, and continuing vertically to the closet ceiling parallel to the walls at a horizontal distance of 300 mm (12 in.) or the width of the shelf, whichever is greater; for a closet that permits access to both sides of a hanging rod, this space includes the volume below the highest rod extending 300 mm (12 in.) on either side of the rod on a plane horizontal to the floor extending the entire length of the rod. See Figure 410.2.

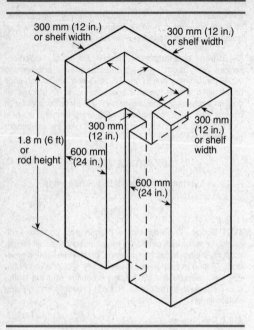

300 mm (12 in.)
or shelf width

300 mm (12 in.)
or shelf width

300 mm
(12 in.) or shelf
width

1.8 m (6 ft)
or rod height

300 mm
(12 in.)

600 mm
(24 in.)

600 mm
(24 in.)

Figure 410.2 Closet Storage Space.

Lighting Track. A manufactured assembly designed to support and energize luminaires that are capable of being readily repositioned on the track. Its length can be altered by the addition or subtraction of sections of track.

410.6 Listing Required. All luminaires and lampholders shall be listed.

410.8 Inspection. Luminaires shall be installed such that the connections between the luminaire conductors and the circuit conductors can be inspected without requiring the disconnection of any part of the wiring unless the luminaires are connected by attachment plugs and receptacles.

II. LUMINAIRE LOCATIONS

410.10 Luminaires in Specific Locations.

(A) Wet and Damp Locations. Luminaires installed in wet or damp locations shall be installed such that water cannot enter or accumulate in wiring compartments, lampholders, or other electrical parts. All luminaires installed in wet locations shall be marked, "Suitable for Wet Locations." All luminaires installed in damp locations shall be marked "Suitable for Wet Locations" or "Suitable for Damp Locations."

(D) Bathtub and Shower Areas. No parts of cord-connected luminaires, chain-, cable-, or cord-suspended luminaires, lighting track, pendants, or ceiling-suspended (paddle) fans shall be located within a zone measured 900 mm (3 ft) horizontally and 2.5 m (8 ft) vertically from the top of the bathtub rim or shower stall threshold. This zone is all encompassing and includes the space directly over the tub or shower stall. Luminaires located within the actual outside dimension of the bathtub or shower to a height of 2.5 m (8 ft) vertically from the top of the bathtub rim or shower threshold shall be marked for damp locations, or marked for wet locations where subject to shower spray.

410.11 Luminaires Near Combustible Material.
Luminaires shall be constructed, installed, or equipped with shades or guards so that combustible material is not subjected to temperatures in excess of 90°C (194°F).

410.12 Luminaires over Combustible Material.
Lampholders installed over highly combustible material shall be of the unswitched type. Unless an individual switch is provided for each luminaire, lampholders shall be located at least 2.5 m (8 ft) above the floor or shall be located or guarded so that the lamps cannot be readily removed or damaged.

410.16 Luminaires in Clothes Closets.

(A) Luminaire Types Permitted. Only luminaires of the following types shall be permitted in a closet:

(1) Surface-mounted or recessed incandescent or LED luminaires with completely enclosed light sources
(2) Surface-mounted or recessed fluorescent luminaires
(3) Surface-mounted fluorescent or LED luminaires identified as suitable for installation within the closet storage space

(B) Luminaire Types Not Permitted. Incandescent luminaires with open or partially enclosed lamps and pendant luminaires or lampholders shall not be permitted.

(C) Location. The minimum clearance between luminaires installed in clothes closets and the nearest point of a closet storage space shall be as follows:

(1) 300 mm (12 in.) for surface-mounted incandescent or LED luminaires with a completely enclosed light source installed on the wall above the door or on the ceiling.
(2) 150 mm (6 in.) for surface-mounted fluorescent luminaires installed on the wall above the door or on the ceiling.
(3) 150 mm (6 in.) for recessed incandescent or LED luminaires with a completely enclosed light source installed in the wall or the ceiling.
(4) 150 mm (6 in.) for recessed fluorescent luminaires installed in the wall or the ceiling.
(5) Surface-mounted fluorescent or LED luminaires shall be permitted to be installed within the closet storage space where identified for this use.

410.18 Space for Cove Lighting. Coves shall have adequate space and shall be located so that lamps and equipment can be properly installed and maintained.

III. PROVISIONS AT LUMINAIRE OUTLET BOXES, CANOPIES, AND PANS

410.22 Outlet Boxes to Be Covered. In a completed installation, each outlet box shall be provided with a cover unless covered by means of a luminaire canopy, lampholder, receptacle, or similar device.

410.23 Covering of Combustible Material at Outlet Boxes. Any combustible wall or ceiling finish exposed between the edge of a luminaire canopy or pan and an outlet box shall be covered with noncombustible material.

410.24 Connection of Electric-Discharge and LED Luminaires.

(A) Independent of the Outlet Box. Electric-discharge and LED luminaires supported independently of the outlet box shall be connected to the branch circuit through metal raceway, nonmetallic raceway, Type MC cable, Type AC cable, Type MI cable, nonmetallic sheathed cable, or by flexible cord as permitted in 410.62(B) or 410.62(C).

(B) Access to Boxes. Electric-discharge and LED luminaires surface mounted over concealed outlet, pull, or junction boxes and designed not to be supported solely by the outlet box shall be provided with suitable openings in the back of the luminaire to provide access to the wiring in the box.

IV. LUMINAIRE SUPPORTS

410.30 Supports.

(A) General. Luminaires and lampholders shall be securely supported. A luminaire that weighs more than 3 kg (6 lb) or exceeds 400 mm (16 in.) in any dimension shall not be supported by the screw shell of a lampholder.

(B) Metal or Nonmetallic Poles Supporting Luminaires. Metal or nonmetallic poles shall be permitted to be used to support luminaires and as a raceway to enclose supply conductors, provided the following conditions are met:

(1) A pole shall have a handhole not less than 50 mm × 100 mm (2 in. × 4 in.) with a cover suitable for use in wet locations to provide access to the supply terminations within the pole or pole base.

Exception No. 1: No handhole shall be required in a pole 2.5 m (8 ft) or less in height abovegrade where the supply wiring method continues without splice or pull point, and where the interior of the pole and any splices are accessible by removing the luminaire.

Exception No. 2: No handhole shall be required in a pole 6.0 m (20 ft) or less in height abovegrade that is provided with a hinged base.

(2) Where raceway risers or cable is not installed within the pole, a threaded fitting or nipple shall be brazed, welded, or attached to the pole opposite the handhole for the supply connection.
(3) A metal pole shall be provided with an equipment grounding terminal as follows:
    a. A pole with a handhole shall have the equipment grounding terminal accessible from the handhole.
    b. A pole with a hinged base shall have the equipment grounding terminal accessible within the base.

Exception to (3): No grounding terminal shall be required in a pole 2.5 m (8 ft) or less in height abovegrade where the supply wiring method continues without splice or pull, and where the interior of the pole and any splices are accessible by removing the luminaire.

(4) A metal pole with a hinged base shall have the hinged base and pole bonded together.
(5) Metal raceways or other equipment grounding conductors shall be bonded to the metal pole with an equipment grounding conductor recognized by 250.118 and sized in accordance with 250.122.
(6) Conductors in vertical poles used as raceway shall be supported as provided in 300.19.

410.36 Means of Support.

(A) Outlet Boxes. Outlet boxes or fittings installed as required by 314.23 and complying with the provisions of 314.27(A)(1) and 314.27(A)(A)(2) shall be permitted to support luminaires.

(B) Suspended Ceilings. Framing members of suspended ceiling systems used to support luminaires shall be securely fastened to each other and shall be securely attached to the building structure at appropriate intervals. Luminaires shall be securely fastened to the ceiling framing member by mechanical means such as bolts, screws, or rivets. Listed clips identified for use with the type of ceiling framing member(s) and luminaire(s) shall also be permitted.

(C) Luminaire Studs. Luminaire studs that are not a part of outlet boxes, hickeys, tripods, and crowfeet shall be made of steel, malleable iron, or other material suitable for the application.

(D) Insulating Joints. Insulating joints that are not designed to be mounted with screws or bolts shall have an exterior metal casing, insulated from both screw connections.

(E) Raceway Fittings. Raceway fittings used to support a luminaire(s) shall be capable of supporting the weight of the complete fixture assembly and lamp(s).

(G) Trees. Outdoor luminaires and associated equipment shall be permitted to be supported by trees.

[**225.26 Vegetation as Support.** Vegetation such as trees shall not be used for support of overhead conductor spans.]

> Informational Note No. 2: See 300.5(D) for protection of conductors.

V. GROUNDING

410.40 General. Luminaires and lighting equipment shall be grounded as required in Article 250 and Part V of this article.

410.42 Luminaire(s) with Exposed Conductive Parts. Exposed metal parts shall be connected to an equipment grounding conductor or insulated from the equipment grounding conductor and other conducting surfaces or be inaccessible to unqualified personnel. Lamp tie wires, mounting screws, clips, and decorative bands on glass spaced at least 38 mm (1-1/2 in.) from lamp terminals shall not be required to be grounded.

410.44 Methods of Grounding. Luminaires and equipment shall be mechanically connected to an equipment grounding conductor as specified in 250.118 and sized in accordance with 250.122.

Exception No. 1: Luminaires made of insulating material that is directly wired or attached to outlets supplied by a wiring method that does not provide a ready means for grounding attachment to an equipment grounding conductor shall be made of insulating material and shall have no exposed conductive parts.

Exception No. 2: Replacement luminaires shall be permitted to connect an equipment grounding conductor from the outlet in compliance with 250.130(C). The luminaire shall then comply with 410.42.

Exception No. 3: Where no equipment grounding conductor exists at the outlet, replacement luminaires that are GFCI protected shall not be required to be connected to an equipment grounding conductor.

VI. WIRING OF LUMINAIRES

410.48 Luminaire Wiring—General. Wiring on or within luminaires shall be neatly arranged and shall not be exposed to physical damage. Excess wiring shall be avoided. Conductors shall be arranged so that they are not subjected to temperatures above those for which they are rated.

410.50 Polarization of Luminaires. Luminaires shall be wired so that the screw shells of lampholders are connected to the same luminaire or circuit conductor or terminal. The grounded conductor, where connected to a screw shell lampholder, shall be connected to the screw shell.

410.56 Protection of Conductors and Insulation.

(A) Properly Secured. Conductors shall be secured in a manner that does not tend to cut or abrade the insulation.

(B) Protection Through Metal. Conductor insulation shall be protected from abrasion where it passes through metal.

(C) Luminaire Stems. Splices and taps shall not be located within luminaire arms or stems.

(D) Splices and Taps. No unnecessary splices or taps shall be made within or on a luminaire.

410.64 Luminaires as Raceways. Luminaires shall not be used as a raceway for circuit conductors unless they comply with 410.64(A), (B), or (C).

(A) Listed. Luminaires listed and marked for use as a raceway shall be permitted to be used as a raceway.

(B) Through-Wiring. Luminaires identified for through-wiring, as permitted by 410.21, shall be permitted to be used as a raceway.

(C) Luminaires Connected Together. Luminaires designed for end-to-end connection to form a continuous assembly, or luminaires connected together by recognized wiring methods, shall be permitted to contain the conductors of a 2-wire branch circuit, or one multiwire branch circuit, supplying the connected luminaires and shall not be required to be listed as a raceway. One additional 2-wire branch circuit separately supplying one or more of the connected luminaires shall also be permitted.

410.68 Feeder and Branch-Circuit Conductors and Ballasts. Feeder and branch-circuit conductors within 75 mm (3 in.) of a ballast, LED driver, power supply, or transformer shall have an insulation temperature rating not lower than 90°C (194°F), unless supplying a luminaire marked as suitable for a different insulation temperature.

X. SPECIAL PROVISIONS FOR FLUSH AND RECESSED LUMINAIRES

410.110 General. Luminaires installed in recessed cavities in walls or ceilings, including suspended ceilings, shall comply with 410.115 through 410.122.

410.115 Temperature.

(A) Combustible Material. Luminaires shall be installed so that adjacent combustible material will not be subjected to temperatures in excess of 90°C (194°F).

(B) Fire-Resistant Construction. Where a luminaire is recessed in fire-resistant material in a building of fire-resistant construction, a temperature higher than 90°C (194°F) but not higher than 150°C (302°F) shall be considered acceptable if the luminaire is plainly marked for that service.

(C) Recessed Incandescent Luminaires. Incandescent luminaires shall have thermal protection and shall be identified as thermally protected.

Exception No. 1: Thermal protection shall not be required in a recessed luminaire identified for use and installed in poured concrete.

Exception No. 2: Thermal protection shall not be required in a recessed luminaire whose design, construction, and thermal performance characteristics are equivalent to a thermally protected luminaire and are identified as inherently protected.

410.116 Clearance and Installation.

(A) Clearance.

(1) Non-Type IC. A recessed luminaire that is not identified for contact with insulation shall have all recessed parts spaced not less than 13 mm (1/2 in.) from combustible materials. The points of support and the trim finishing off the openings in the ceiling, wall, or other finished surface shall be permitted to be in contact with combustible materials.

(2) Type IC. A recessed luminaire that is identified for contact with insulation, Type IC, shall be permitted to be in contact with combustible materials at recessed parts, points of support, and portions passing through or finishing off the opening in the building structure.

(B) Installation. Thermal insulation shall not be installed above a recessed luminaire or within 75 mm (3 in.) of the recessed luminaire's enclosure, wiring compartment, ballast, transformer, LED driver, or power supply unless the luminaire is identified as Type IC for insulation contact.

410.117 Wiring.

(A) General. Conductors that have insulation suitable for the temperature encountered shall be used.

(B) Circuit Conductors. Branch-circuit conductors that have an insulation suitable for the temperature encountered shall be permitted to terminate in the luminaire.

(C) Tap Conductors. Tap conductors of a type suitable for the temperature encountered shall be permitted to run from the luminaire terminal connection to an outlet box placed at least 300 mm (1 ft) from the luminaire. Such tap conductors shall be in suitable raceway or Type AC or MC cable of at least 450 mm (18 in.) but not more than 1.8 m (6 ft) in length.

XIV. LIGHTING TRACK

410.151 Installation.

(A) Lighting Track. Lighting track shall be permanently installed and permanently connected to a branch circuit. Only lighting track fittings shall be installed on lighting track. Lighting track fittings shall not be equipped with general-purpose receptacles.

(B) Connected Load. The connected load on lighting track shall not exceed the rating of the track. Lighting track shall be supplied by a branch circuit having a rating not more than that of the track.

(C) Locations Not Permitted. Lighting track shall not be installed in the following locations:

(1) Where likely to be subjected to physical damage
(2) In wet or damp locations
(6) Where concealed
(7) Where extended through walls or partitions
(8) Less than 1.5 m (5 ft) above the finished floor except where protected from physical damage or track operating at less than 30 volts rms open-circuit voltage
(9) Where prohibited by 410.10(D)

(D) Support. Fittings identified for use on lighting track shall be designed specifically for the track on which they are to be installed. They shall be securely fastened to the track, shall maintain polarization and connections to the equipment grounding conductor, and shall be designed to be suspended directly from the track.

410.154 Fastening. Lighting track shall be securely mounted so that each fastening is suitable for supporting the maximum weight of luminaires that can be installed. Unless identified for supports at greater intervals, a single section 1.2 m (4 ft) or shorter in length shall have two supports, and, where installed in a continuous row, each individual section of not more than 1.2 m (4 ft) in length shall have one additional support.

410.155 Construction Requirements.

(B) Grounding. Lighting track shall be grounded in accordance with Article 250, and the track sections shall be securely coupled to maintain continuity of the circuitry, polarization, and grounding throughout.

ARTICLE 411
Lighting Systems Operating at 30 Volts or Less

411.1 Scope. This article covers lighting systems operating at 30 volts or less and their associated components.

411.2 Definition.

Lighting Systems Operating at 30 Volts or Less. A lighting system consisting of an isolating power supply, the low-voltage luminaires, and associated equipment that are all identified for the use. The output circuits of the power supply are rated for not more than 25 amperes and operate at 30 volts (42.4 volts peak) or less under all load conditions.

411.3 Listing Required. Lighting systems operating at 30 volts or less shall comply with 411.3(A) or 411.3(B).

(A) Listed System. Lighting systems operating at 30 volts or less shall be listed as a complete system. The luminaires, power supply, and luminaire fittings (including the exposed bare conductors) of an exposed bare conductor lighting system shall be listed for the use as part of the same identified lighting system.

(B) Assembly of Listed Parts. A lighting system assembled from the following listed parts shall be permitted:

(1) Low-voltage luminaires
(2) Low-voltage luminaire power supply
(3) Class 2 power supply
(4) Low-voltage luminaire fittings

(5) Cord (secondary circuit) for which the luminaires and power supply are listed for use
(6) Cable, conductors in conduit, or other fixed wiring method for the secondary circuit

The luminaires, power supply, and luminaire fittings (including the exposed bare conductors) of an exposed bare conductor lighting system shall be listed for use as part of the same identified lighting system.

411.4 Specific Location Requirements.

(A) Walls, Floors, and Ceilings. Conductors concealed or extended through a wall, floor, or ceiling shall be in accordance with (1) or (2):

(1) Installed using any of the wiring methods specified in Chapter 3
(2) Installed using wiring supplied by a listed Class 2 power source and installed in accordance with 725.130

(B) Pools, Spas, Fountains, and Similar Locations. Lighting systems shall be installed not less than 3 m (10 ft) horizontally from the nearest edge of the water, unless permitted by Article 680.

411.5 Secondary Circuits.

(A) Grounding. Secondary circuits shall not be grounded.

(B) Isolation. The secondary circuit shall be insulated from the branch circuit by an isolating transformer.

(C) Bare Conductors. Exposed bare conductors and current-carrying parts shall be permitted for indoor installations only. Bare conductors shall not be installed less than 2.1 m (7 ft) above the finished floor, unless specifically listed for a lower installation height.

(D) Insulated Conductors. Exposed insulated secondary circuit conductors shall be of the type, and installed as, described in (1), (2), or (3):

(1) Class 2 cable supplied by a Class 2 power source and installed in accordance with Parts I and III of Article 725.

(2) Conductors, cord, or cable of the listed system and installed not less than 2.1 m (7 ft) above the finished floor unless the system is specifically listed for a lower installation height.

(3) Wiring methods described in Chapter 3

411.6 Branch Circuit. Lighting systems operating at 30 volts or less shall be supplied from a maximum 20-ampere branch circuit.

CHAPTER 20

APPLIANCES

INTRODUCTION

This chapter contains provisions from Article 422—Appliances. Article 422 covers electric appliances used in any occupancy. Installation provisions cover branch-circuit ratings and overcurrent protection. Flexible-cord requirements are provided for electrically operated kitchen waste disposers; built-in dishwashers and trash compactors; and wall-mounted ovens and counter-mounted cooking units. Appliance disconnecting specifications are covered in Part III of Article 422.

ARTICLE 422
Appliances

I. GENERAL

422.1 Scope. This article covers electrical appliances used in any occupancy.

II. INSTALLATION

422.10 Branch-Circuit Rating. This section specifies the ratings of branch circuits capable of carrying appliance current without overheating under the conditions specified.

(A) Individual Circuits. The rating of an individual branch circuit shall not be less than the marked rating of the appliance or the marked rating of an appliance having combined loads as provided in 422.62.

The rating of an individual branch circuit for motor-operated appliances not having a marked rating shall be in accordance with Part II of Article 430.

The branch-circuit rating for an appliance that is a continuous load, other than a motor-operated appliance, shall not be less than 125 percent of the marked rating, or not less than 100 percent of the marked rating if the branch-circuit device and its assembly are listed for continuous loading at 100 percent of its rating.

Branch circuits and branch-circuit conductors for household ranges and cooking appliances shall be permitted to be in accordance with Table 220.55 and shall be sized in accordance with 210.19(A)(3).

(B) Circuits Supplying Two or More Loads. For branch circuits supplying appliance and other loads, the rating shall be determined in accordance with 210.23.

422.11 Overcurrent Protection. Appliances shall be protected against overcurrent in accordance with 422.11(A) through (G) and 422.10.

(A) Branch-Circuit Overcurrent Protection. Branch circuits shall be protected in accordance with 240.4.

If a protective device rating is marked on an appliance, the branch-circuit overcurrent device rating shall not exceed the protective device rating marked on the appliance.

(B) Household-Type Appliances with Surface Heating Elements. Household-type appliances with surface heating elements having a maximum demand of more than 60 amperes calculated in accordance with Table 220.55 shall have their power supply subdivided into two or more circuits, each of which shall be provided with overcurrent protection rated at not over 50 amperes.

(E) Single Non–motor-Operated Appliance. If the branch circuit supplies a single non–motor-operated appliance, the rating of overcurrent protection shall comply with the following:

(1) Not exceed that marked on the appliance.

(2) Not exceed 20 amperes if the overcurrent protection rating is not marked and the appliance is rated 13.3 amperes or less; or

(3) Not exceed 150 percent of the appliance rated current if the overcurrent protection rating is not marked and the appliance is rated over 13.3 amperes. Where 150 percent of the appliance rating does not correspond to a standard overcurrent device ampere rating, the next higher standard rating shall be permitted.

(G) Motor-Operated Appliances. Motors of motor-operated appliances shall be provided with overload protection in accordance with Part III of Article 430. Hermetic refrigerant motor-compressors in air-conditioning or refrigerating equipment shall be provided with overload protection in accordance with Part VI of Article 440. Where appliance overcurrent protective devices that are separate from the appliance are required, data for selection of these devices shall be marked on the appliance. The minimum marking shall be that specified in 430.7 and 440.4.

422.12 Central Heating Equipment. Central heating equipment other than fixed electric space-heating equipment shall be supplied by an individual branch circuit.

Exception No. 1: Auxiliary equipment, such as a pump, valve, humidifier, or electrostatic air cleaner directly associated with the heating equipment, shall be permitted to be connected to the same branch circuit.

Exception No. 2: Permanently connected air-conditioning equipment shall be permitted to be connected to the same branch circuit.

422.13 Storage-Type Water Heaters. A fixed storage-type water heater that has a capacity of 450 L (120 gal) or less shall be considered a continuous load for the purposes of sizing branch circuits.

422.15 Central Vacuum Outlet Assemblies.

(A) Listed central vacuum outlet assemblies shall be permitted to be connected to a branch circuit in accordance with 210.23(A).

(B) The ampacity of the connecting conductors shall not be less than the ampacity of the branch circuit conductors to which they are connected.

(C) Accessible non–current-carrying metal parts of the central vacuum outlet assembly likely to become energized shall be connected to an equipment grounding conductor in accordance with 250.110. Incidental metal parts such as screws or rivets installed into or on insulating material shall not be considered likely to become energized.

422.16 Flexible Cords.

(A) General. Flexible cord shall be permitted (1) for the connection of appliances to facilitate their frequent interchange or to prevent the transmission of noise or vibration or (2) to facilitate the removal or disconnection of appliances that are fastened in place, where the fastening means and mechanical connections are specifically designed to permit ready removal for maintenance or repair and the appliance is intended or identified for flexible cord connection.

(B) Specific Appliances.

(1) Electrically Operated Kitchen Waste Disposers. Electrically operated kitchen waste disposers shall be permitted to be cord-and-plug-connected with a flexible cord identified as suitable for the purpose in the installation instructions of the appliance manufacturer, where all of the following conditions are met:

(1) The flexible cord shall be terminated with a grounding-type attachment plug.

Exception: A listed kitchen waste disposer distinctly marked to identify it as protected by a system of double insulation, or its equivalent, shall not be required to be terminated with a grounding-type attachment plug.

(2) The length of the cord shall not be less than 450 mm (18 in.) and not over 900 mm (36 in.).

(3) Receptacles shall be located to avoid physical damage to the flexible cord.

(4) The receptacle shall be accessible.

(2) Built-in Dishwashers and Trash Compactors. Built-in dishwashers and trash compactors shall be permitted to be cord-and-plug-connected with a flexible cord identified as suitable for the purpose in the installation instructions of the appliance manufacturer where all of the following conditions are met:

(1) The flexible cord shall be terminated with a grounding-type attachment plug.

Exception: A listed dishwasher or trash compactor distinctly marked to identify it as protected by a system of double insulation, or its equivalent, shall not be required to be terminated with a grounding-type attachment plug.

(2) The length of the cord shall be 0.9 m to 1.2 m (3 ft to 4 ft) measured from the face of the attachment plug to the plane of the rear of the appliance.

(3) Receptacles shall be located to avoid physical damage to the flexible cord.

(4) The receptacle shall be located in the space occupied by the appliance or adjacent thereto.

(5) The receptacle shall be accessible.

(3) Wall-Mounted Ovens and Counter-Mounted Cooking Units. Wall-mounted ovens and counter-mounted cooking units complete with provisions for mounting and for making electrical connections shall be permitted to be permanently connected or, only for ease in servicing or for installation, cord-and-plug-connected.

A separable connector or a plug and receptacle combination in the supply line to an oven or cooking unit shall be approved for the temperature of the space in which it is located.

(4) Range Hoods. Range hoods shall be permitted to be cord-and-plug-connected with a flexible cord identified as suitable for use on range hoods in the installation instructions of the appliance manufacturer, where all of the following conditions are met:

(1) The flexible cord is terminated with a grounding-type attachment plug.

Exception: A listed range hood distinctly marked to identify it as protected by a system of double insulation, or its equivalent, shall not be required to be terminated with a grounding-type attachment plug.

(2) The length of the cord is not less than 450 mm (18 in.) and not over 900 mm (36 in.).
(3) Receptacles are located to avoid physical damage to the flexible cord.
(4) The receptacle is accessible.
(5) The receptacle is supplied by an individual branch circuit.

422.18 Support of Ceiling-Suspended (Paddle) Fans. Ceiling-suspended (paddle) fans shall be supported independently of an outlet box or by listed outlet box or outlet box systems identified for the use and installed in accordance with 314.27(C).

422.20 Other Installation Methods. Appliances employing methods of installation other than covered by this article shall be permitted to be used only by special permission.

III. DISCONNECTING MEANS

422.30 General. A means shall be provided to simultaneously disconnect each appliance from all ungrounded conductors in accordance with the following sections of Part III. If an appliance is supplied by more than one branch-circuit or feeder, these disconnecting means shall be grouped and identified as the appliance disconnect.

422.31 Disconnection of Permanently Connected Appliances.

(A) Rated at Not over 300 Volt-Amperes or 1/8 Horsepower. For permanently connected appliances rated at not over 300 volt-amperes or 1/8 hp, the branch-circuit overcurrent device shall be permitted to serve as the disconnecting means.

(B) Appliances Rated over 300 Volt-Amperes. For permanently connected appliances rated over 300 volt-amperes, the branch-circuit switch or circuit breaker shall be permitted to serve as the disconnecting means where the switch or circuit breaker is within sight from the appliance or is capable of being locked in the open position. The provision for locking or adding a lock to the disconnecting means shall be installed on or at the switch or circuit breaker used as the disconnecting means and shall remain in place with or without the lock installed.

(C) Motor-Operated Appliances Rated over 1/8 Horsepower. For permanently connected motor-operated appliances with motors rated over 1/8 horse power, the branch-circuit switch or circuit breaker shall be permitted to serve as the disconnecting means where the switch or circuit breaker is within sight from the appliance. The disconnecting means shall comply with 430.109 and 430.110.

Exception: If an appliance of more than 1/8 hp is provided with a unit switch that complies with 422.34(A), (B), (C), or (D), the switch or circuit breaker serving as the other disconnecting means shall be permitted to be out of sight from the appliance.

422.33 Disconnection of Cord-and-Plug-Connected Appliances.

(A) Separable Connector or an Attachment Plug and Receptacle. For cord-and-plug-connected appliances, an accessible separable connector or an accessible plug and receptacle shall be permitted to serve as the disconnecting means. Where the separable connector or plug and receptacle are not accessible, cord-and-plug-connected appliances shall be provided with disconnecting means in accordance with 422.31.

(B) Connection at the Rear Base of a Range. For cord-and-plug-connected household electric ranges, an attachment plug and receptacle connection at the rear base of a range, if it is accessible from the front by removal of a drawer, shall be considered as meeting the intent of 422.33(A).

(C) Rating. The rating of a receptacle or of a separable connector shall not be less than the rating of any appliance connected thereto.

422.34 Unit Switch(es) as Disconnecting Means. A unit switch(es) with a marked-off position that is a part of an appliance and disconnects all ungrounded conductors shall be permitted as the disconnecting means required by this article where other means for disconnection are provided in occupancies specified in 422.34(A) through (D).

(A) Multifamily Dwellings. In multifamily dwellings, the other disconnecting means shall be within the dwelling unit, or on the same floor as the dwelling unit in which the appliance is installed, and shall be permitted to control lamps and other appliances.

(B) Two-Family Dwellings. In two-family dwellings, the other disconnecting means shall be permitted either inside or outside of the dwelling unit in which the appliance is installed. In this case, an individual switch or circuit breaker for the dwelling unit shall be permitted and shall also be permitted to control lamps and other appliances.

(C) One-Family Dwellings. In one-family dwellings, the service disconnecting means shall be permitted to be the other disconnecting means.

422.35 Switch and Circuit Breaker to Be Indicating. Switches and circuit breakers used as disconnecting means shall be of the indicating type.

CHAPTER 21

ELECTRIC SPACE-HEATING EQUIPMENT

INTRODUCTION

This chapter contains provisions from Article 424—Fixed Electric Space-Heating Equipment. Article 424 covers fixed electric equipment used for space heating. Heating equipment covered in this article includes: heating cable, unit heaters, central systems, and other approved fixed electric space-heating equipment. Contained in this pocket guide are four of the nine parts that are included in Article 424. The four parts are: I General; II Installation; III Control and Protection of Fixed Electric Space-Heating Equipment; and V Electric Space-Heating Cables. Room air-conditioning equipment is covered in Article 440 (see Chapter 22).

ARTICLE 424
Fixed Electric Space-Heating Equipment

I. GENERAL

424.1 Scope. This article covers fixed electric equipment used for space heating. For the purpose of this article, heating equipment shall include heating cable, unit heaters, boilers, central systems, or other approved fixed electric space-heating equipment. This article shall not apply to process heating and room air conditioning.

424.3 Branch Circuits.

(A) Branch-Circuit Requirements. Individual branch circuits shall be permitted to supply any volt-ampere or wattage rating of fixed electric space-heating equipment for which they are rated.

Branch circuits supplying two or more outlets for fixed electric space-heating equipment shall be rated 15, 20, 25, or 30 amperes. In other than a dwelling unit, fixed infrared heating equipment shall be permitted to be supplied from branch circuits rated not over 50 amperes.

(B) Branch-Circuit Sizing. Fixed electric space-heating equipment and motors shall be considered continuous load.

424.6 Listed Equipment. Electric baseboard heaters, heating cables, duct heaters, and radiant heating systems shall be listed and labeled.

II. INSTALLATION

424.9 General. All fixed electric space-heating equipment shall be installed in an approved manner.

Permanently installed electric baseboard heaters equipped with factory-installed receptacle outlets, or outlets provided as a separate listed assembly, shall be permitted in lieu of a receptacle outlet(s) that is required by 210.50(B). Such receptacle outlets shall not be connected to the heater circuits.

Informational Note: Listed baseboard heaters include instructions that may not permit their installation below receptacle outlets.

424.12 Locations.

(A) Exposed to Physical Damage. Where subject to physical damage, fixed electric space-heating equipment shall be protected in an approved manner.

(B) Damp or Wet Locations. Heaters and related equipment installed in damp or wet locations shall be listed for such locations and shall be constructed and installed so that water or other liquids cannot enter or accumulate in or on wired sections, electrical components, or ductwork.

Informational Note No. 1: See 110.11 for equipment exposed to deteriorating agents.

Informational Note No. 2: See 680.27(C) for pool deck areas.

424.13 Spacing from Combustible Materials. Fixed electric space-heating equipment shall be installed to provide the required spacing between the equipment and adjacent combustible material, unless it is listed to be installed in direct contact with combustible material.

III. CONTROL AND PROTECTION OF FIXED ELECTRIC SPACE-HEATING EQUIPMENT

424.19 Disconnecting Means. Means shall be provided to simultaneously disconnect the heater, motor controller(s), and supplementary overcurrent protective device(s) of all fixed electric space-heating equipment from all ungrounded conductors. Where heating equipment is supplied by more than one source, the disconnecting means shall be grouped and marked. The disconnecting means specified in 424.19(A) and (B) shall have an ampere rating not less than 125 percent of the total load of the motors and the heaters. The provision for locking or adding a lock to the disconnecting means shall be installed on or at the switch or circuit breaker used as the disconnecting means and shall remain in place with or without the lock installed.

(A) Heating Equipment with Supplementary Overcurrent Protection. The disconnecting means for fixed electric space-heating equipment with supplementary overcurrent protection shall be within sight from the supplementary overcurrent protective device(s), on the supply side of these devices, if fuses, and, in addition, shall comply with either 424.19(A)(1) or (A)(2).

(1) Heater Containing No Motor Rated over 1/8 Horsepower. The above disconnecting means or unit switches complying with 424.19(C) shall be permitted to serve as the required disconnecting means for both the motor controller(s) and heater under either of the following conditions:

(1) The disconnecting means provided is also within sight from the motor controller(s) and the heater.
(2) The disconnecting means provided is capable of being locked in the open (off) position.

(2) Heater Containing a Motor(s) Rated over 1/8 Horsepower. The above disconnecting means shall be permitted to serve as the required disconnecting means for both the motor controller(s) and heater under either of the following conditions:

(1) Where the disconnecting means is in sight from the motor controller(s) and the heater and complies with Part IX of Article 430.
(2) Where a motor(s) of more than 1/8 hp and the heater are provided with a single unit switch that complies with 422.34(A), (B), (C), or (D), the disconnecting means shall be permitted to be out of sight from the motor controller.

(B) Heating Equipment Without Supplementary Overcurrent Protection.

(1) Without Motor or with Motor Not over 1/8 Horsepower. For fixed electric space-heating equipment without a motor rated over 1/8 hp, the branch-circuit switch or circuit breaker shall be permitted to serve as the disconnecting means where the switch or circuit breaker is within sight from the heater or is capable of being locked in the open (off) position.

(2) Over 1/8 Horsepower. For motor-driven electric space-heating equipment with a motor rated over 1/8 hp, a disconnecting means shall be located within sight from the motor controller or shall be permitted to comply with the requirements in 424.19(A)(2).

(C) Unit Switch(es) as Disconnecting Means. A unit switch(es) with a marked "off" position that is part of a fixed heater and disconnects all ungrounded conductors shall be permitted as the disconnecting means required by this article where other means for disconnection are provided in the types of occupancies in 424.19(C)(1) through (C)(4).

(1) Multifamily Dwellings. In multifamily dwellings, the other disconnecting means shall be within the dwelling unit, or on the same floor as the dwelling unit in which the fixed heater is installed, and shall also be permitted to control lamps and appliances.

(2) Two-Family Dwellings. In two-family dwellings, the other disconnecting means shall be permitted either inside or outside of the dwelling unit in which the fixed heater is installed. In this case, an individual switch or circuit breaker for the dwelling unit shall be permitted and shall also be permitted to control lamps and appliances.

(3) One-Family Dwellings. In one-family dwellings, the service disconnecting means shall be permitted to be the other disconnecting means.

424.20 Thermostatically Controlled Switching Devices.

(A) Serving as Both Controllers and Disconnecting Means. Thermostatically controlled switching devices and combination thermostats and manually controlled switches shall be permitted to serve as both controllers and disconnecting means, provided they meet all of the following conditions:

(1) Provided with a marked "off" position
(2) Directly open all ungrounded conductors when manually placed in the "off" position

(3) Designed so that the circuit cannot be energized automatically after the device has been manually placed in the "off" position

(4) Located as specified in 424.19

(B) Thermostats That Do Not Directly Interrupt All Ungrounded Conductors. Thermostats that do not directly interrupt all ungrounded conductors and thermostats that operate remote-control circuits shall not be required to meet the requirements of 424.20(A). These devices shall not be permitted as the disconnecting means.

424.21 Switch and Circuit Breaker to Be Indicating. Switches and circuit breakers used as disconnecting means shall be of the indicating type.

424.22 Overcurrent Protection.

(A) Branch-Circuit Devices. Electric space-heating equipment, other than such motor-operated equipment as required by Articles 430 and 440 to have additional overcurrent protection, shall be permitted to be protected against overcurrent where supplied by one of the branch circuits in Article 210.

V. ELECTRIC SPACE-HEATING CABLES

424.35 Marking of Heating Cables. Each unit shall be marked with the identifying name or identification symbol, catalog number, and ratings in volts and watts or in volts and amperes.

Each unit length of heating cable shall have a permanent legible marking on each nonheating lead located within 75 mm (3 in.) of the terminal end. The lead wire shall have the following color identification to indicate the circuit voltage on which it is to be used:

(1) 120 volt, nominal—yellow
(2) 208 volt, nominal—blue
(3) 240 volt, nominal—red
(4) 277 volt, nominal—brown
(5) 480 volt, nominal—orange

424.36 Clearances of Wiring in Ceilings. Wiring located above heated ceilings shall be spaced not less than 50 mm

(2 in.) above the heated ceiling and shall be considered as operating at an ambient temperature of 50°C (122°F). The ampacity of conductors shall be calculated on the basis of the correction factors shown in the 0–2000 volt ampacity tables of Article 310. If this wiring is located above thermal insulation having a minimum thickness of 50 mm (2 in.), the wiring shall not require correction for temperature.

424.38 Area Restrictions.

(A) Shall Not Extend Beyond the Room or Area. Heating cables shall not extend beyond the room or area in which they originate.

(B) Uses Prohibited. Heating cables shall not be installed in the following:

(1) In closets
(2) Over walls
(3) Over partitions that extend to the ceiling, unless they are isolated single runs of embedded cable
(4) Over cabinets whose clearance from the ceiling is less than the minimum horizontal dimension of the cabinet to the nearest cabinet edge that is open to the room or area

(C) In Closet Ceilings as Low-Temperature Heat Sources to Control Relative Humidity. The provisions of 424.38(B) shall not prevent the use of cable in closet ceilings as low-temperature heat sources to control relative humidity, provided they are used only in those portions of the ceiling that are unobstructed to the floor by shelves or other permanent luminaires.

424.39 Clearance from Other Objects and Openings.
Heating elements of cables shall be separated at least 200 mm (8 in.) from the edge of outlet boxes and junction boxes that are to be used for mounting surface luminaires. A clearance of not less than 50 mm (2 in.) shall be provided from recessed luminaires and their trims, ventilating openings, and other such openings in room surfaces. No heating cable shall be covered by any surface-mounted equipment.

424.40 Splices. Embedded cables shall be spliced only where necessary and only by approved means, and in no case shall the length of the heating cable be altered.

424.41 Installation of Heating Cables on Dry Board, in Plaster, and on Concrete Ceilings.

(A) In Walls. Cables shall not be installed in walls unless it is necessary for an isolated single run of cable to be installed down a vertical surface to reach a dropped ceiling.

(B) Adjacent Runs. Adjacent runs of cable not exceeding 9 watts/m (2-3/4 watts/ft) shall not be installed less than 38 mm (1-1/2 in.) on centers.

(C) Surfaces to Be Applied. Heating cables shall be applied only to gypsum board, plaster lath, or other fire-resistant material. With metal lath or other electrically conductive surfaces, a coat of plaster shall be applied to completely separate the metal lath or conductive surface from the cable.

(D) Splices. All heating cables, the splice between the heating cable and nonheating leads, and 75-mm (3-in.) minimum of the nonheating lead at the splice shall be embedded in plaster or dry board in the same manner as the heating cable.

(E) Ceiling Surface. The entire ceiling surface shall have a finish of thermally noninsulating sand plaster that has a nominal thickness of 13 mm (1/2 in.), or other noninsulating material identified as suitable for this use and applied according to specified thickness and directions.

(F) Secured. Cables shall be secured by means of approved stapling, tape, plaster, nonmetallic spreaders, or other approved means either at intervals not exceeding 400 mm (16 in.) or at intervals not exceeding 1.8 m (6 ft) for cables identified for such use. Staples or metal fasteners that straddle the cable shall not be used with metal lath or other electrically conductive surfaces.

(G) Dry Board Installations. In dry board installations, the entire ceiling below the heating cable shall be covered with gypsum board not exceeding 13 mm (1/2 in.) thickness. The void between the upper layer of gypsum board, plaster lath, or other fire-resistant material and the surface layer of gypsum board shall be completely filled with thermally conductive, nonshrinking plaster or other approved material or equivalent thermal conductivity.

(H) Free from Contact with Conductive Surfaces. Cables shall be kept free from contact with metal or other electrically conductive surfaces.

(I) Joists. In dry board applications, cable shall be installed parallel to the joist, leaving a clear space centered under the joist of 65 mm (2-1/2 in.) (width) between centers of adjacent runs of cable. A surface layer of gypsum board shall be mounted so that the nails or other fasteners do not pierce the heating cable.

(J) Crossing Joists. Cables shall cross joists only at the ends of the room unless the cable is required to cross joists elsewhere in order to satisfy the manufacturer's instructions that the installer avoid placing the cable too close to ceiling penetrations and luminaires.

424.42 Finished Ceilings. Finished ceilings shall not be covered with decorative panels or beams constructed of materials that have thermal insulating properties, such as wood, fiber, or plastic. Finished ceilings shall be permitted to be covered with paint, wallpaper, or other approved surface finishes.

424.43 Installation of Nonheating Leads of Cables.

(A) Free Nonheating Leads. Free nonheating leads of cables shall be installed in accordance with approved wiring methods from the junction box to a location within the ceiling. Such installations shall be permitted to be single conductors in approved raceways, single or multiconductor Type UF, Type NMC, Type MI, or other approved conductors.

(B) Leads in Junction Box. Not less than 150 mm (6 in.) of free nonheating lead shall be within the junction box. The marking of the leads shall be visible in the junction box.

(C) Excess Leads. Excess leads of heating cables shall not be cut but shall be secured to the underside of the ceiling and embedded in plaster or other approved material, leaving only a length sufficient to reach the junction box with not less than 150 mm (6 in.) of free lead within the box.

424.44 Installation of Cables in Concrete or Poured Masonry Floors.

(A) Watts per Linear Meter (Foot). Constant wattage heating cables shall not exceed 54 watts per linear meter (16-1/2 watts per linear foot) of cable.

(B) Spacing Between Adjacent Runs. The spacing between adjacent runs of cable shall not be less than 25 mm (1 in.) on centers.

(C) Secured in Place. Cables shall be secured in place by nonmetallic frames or spreaders or other approved means while the concrete or other finish is applied.

Cables shall not be installed where they bridge expansion joints unless protected from expansion and contraction.

(D) Spacings Between Heating Cable and Metal Embedded in the Floor. Spacings shall be maintained between the heating cable and metal embedded in the floor, unless the cable is a grounded metal-clad cable.

(E) Leads Protected. Leads shall be protected where they leave the floor by rigid metal conduit, intermediate metal conduit, rigid nonmetallic conduit, electrical metallic tubing, or by other approved means.

(F) Bushings or Approved Fittings. Bushings or approved fittings shall be used where the leads emerge within the floor slab.

(G) Ground-Fault Circuit-Interrupter Protection. Ground-fault circuit-interrupter protection for personnel shall be provided for cables installed in electrically heated floors of bathrooms, kitchens, and in hydromassage bathtub locations.

424.45 Inspection and Tests. Cable installations shall be made with due care to prevent damage to the cable assembly and shall be inspected and approved before cables are covered or concealed.

CHAPTER 22

AIR-CONDITIONING AND REFRIGERATING EQUIPMENT

INTRODUCTION

This chapter contains provisions from Article 440—Air-Conditioning and Refrigerating Equipment. Article 440 covers electric motor–driven air-conditioning and refrigerating equipment, including their branch circuits and controllers. Provided in Article 440 are special considerations necessary for circuits supplying hermetic refrigerant motor-compressors and for any air-conditioning or refrigerating equipment that is supplied from a branch circuit that supplies a hermetic refrigerant motor-compressor.

Contained in this pocket guide are six of the seven parts that are included in Article 440. The six parts are: I General; II Disconnecting Means; III Branch-Circuit Short-Circuit and Ground-Fault Protection; IV Branch-Circuit Conductors; VI Motor-Compressor and Branch-Circuit Overload Protection; and VII Provisions for Room Air Conditioners.

ARTICLE 440
Air-Conditioning and Refrigerating Equipment

I. GENERAL

440.1 Scope. The provisions of this article apply to electric motor-driven air-conditioning and refrigerating equipment and to the branch circuits and controllers for such equipment. It provides for the special considerations necessary for circuits supplying hermetic refrigerant motor-compressors and for any air-conditioning or refrigerating equipment that is supplied from a branch circuit that supplies a hermetic refrigerant motor-compressor.

440.2 Definitions.

Branch-Circuit Selection Current. The value in amperes to be used instead of the rated-load current in determining the ratings of motor branch-circuit conductors, disconnecting means, controllers, and branch-circuit short-circuit and ground-fault protective devices wherever the running overload protective device permits a sustained current greater than the specified percentage of the rated-load current. The value of branch-circuit selection current will always be equal to or greater than the marked rated-load current.

Hermetic Refrigerant Motor-Compressor. A combination consisting of a compressor and motor, both of which are enclosed in the same housing, with no external shaft or shaft seals, the motor operating in the refrigerant.

Leakage-Current Detector-Interrupter (LCDI). A device provided in a power supply cord or cord set that senses leakage current flowing between or from the cord conductors and interrupts the circuit at a predetermined level of leakage current.

Rated-Load Current. The rated-load current for a hermetic refrigerant motor-compressor is the current resulting when the motor-compressor is operated at the rated load, rated voltage, and rated frequency of the equipment it serves.

440.3 Other Articles.

(A) Article 430. These provisions are in addition to, or amendatory of, the provisions of Article 430 and other articles in this *Code*, which apply except as modified in this article.

(B) Articles 422, 424, or 430. The rules of Articles 422, 424, or 430, as applicable, shall apply to air-conditioning and refrigerating equipment that does not incorporate a hermetic refrigerant motor-compressor. This equipment includes devices that employ refrigeration compressors driven by conventional motors, furnaces with air-conditioning evaporator coils installed, fan-coil units, remote forced air-cooled condensers, remote commercial refrigerators, and so forth.

(C) Article 422. Equipment such as room air conditioners, household refrigerators and freezers, drinking water coolers, and beverage dispensers shall be considered appliances, and the provisions of Article 422 shall also apply.

440.6 Ampacity and Rating. The size of conductors for equipment covered by this article shall be selected from Table 310.15(B)(16) through Table 310.15(B)(19) or calculated in accordance with 310.15 as applicable. The required ampacity of conductors and rating of equipment shall be determined according to 440.6(A) and 440.6(B).

(A) Hermetic Refrigerant Motor-Compressor. For a hermetic refrigerant motor-compressor, the rated-load current marked on the nameplate of the equipment in which the motor-compressor is employed shall be used in determining the rating or ampacity of the disconnecting means, the branch-circuit conductors, the controller, the branch-circuit short-circuit and ground-fault protection, and the separate motor overload protection. Where no rated-load current is shown on the equipment nameplate, the rated-load current shown on the compressor nameplate shall be used.

Exception No. 1: Where so marked, the branch-circuit selection current shall be used instead of the rated-load current to determine the rating or ampacity of the disconnecting means, the branch-circuit conductors, the controller, and the branch-circuit short-circuit and ground-fault protection.

(B) Multimotor Equipment. For multimotor equipment employing a shaded-pole or permanent split-capacitor-type fan or blower motor, the full-load current for such motor marked on the nameplate of the equipment in which the fan or blower motor is employed shall be used instead of the horsepower rating to determine the ampacity or rating of the disconnecting means, the branch-circuit conductors, the controller, the branch-circuit short-circuit and ground-fault protection, and the separate overload protection. This marking on the equipment nameplate shall not be less than the current marked on the fan or blower motor nameplate.

440.8 Single Machine. An air-conditioning or refrigerating system shall be considered to be a single machine under the provisions of 430.87, Exception, and 430.112, Exception. The motors shall be permitted to be located remotely from each other.

II. DISCONNECTING MEANS

440.11 General. The provisions of Part II are intended to require disconnecting means capable of disconnecting air-conditioning and refrigerating equipment, including motor-compressors and controllers from the circuit conductors.

440.12 Rating and Interrupting Capacity.

(A) Hermetic Refrigerant Motor-Compressor. A disconnecting means serving a hermetic refrigerant motor-compressor shall be selected on the basis of the nameplate rated-load current or branch-circuit selection current, whichever is greater, and locked-rotor current, respectively, of the motor-compressor as follows.

(1) Ampere Rating. The ampere rating shall be at least 115 percent of the nameplate rated-load current or branch-circuit selection current, whichever is greater.

Exception: A listed unfused motor circuit switch, without fuseholders, having a horsepower rating not less than the equivalent horsepower determined in accordance with 440.12(A)(2) shall be permitted to have an ampere rating less than 115 percent of the specified current.

440.13 Cord-Connected Equipment. For cord-connected equipment such as room air conditioners, household refrigerators and freezers, drinking water coolers, and beverage dispensers, a separable connector or an attachment plug and receptacle shall be permitted to serve as the disconnecting means.

Informational Note: For room air conditioners, see 440.63.

440.14 Location. Disconnecting means shall be located within sight from and readily accessible from the air-conditioning or refrigerating equipment. The disconnecting means shall be permitted to be installed on or within the air-conditioning or refrigerating equipment.

The disconnecting means shall not be located on panels that are designed to allow access to the air-conditioning or refrigeration equipment or to obscure the equipment nameplate(s).

Exception No. 2: Where an attachment plug and receptacle serve as the disconnecting means in accordance with 440.13, their location shall be accessible but shall not be required to be readily accessible.

III. BRANCH-CIRCUIT SHORT-CIRCUIT AND GROUND-FAULT PROTECTION

440.21 General. The provisions of Part III specify devices intended to protect the branch-circuit conductors, control apparatus, and motors in circuits supplying hermetic refrigerant motor-compressors against overcurrent due to short circuits and ground faults. They are in addition to or amendatory of the provisions of Article 240.

440.22 Application and Selection.

(A) Rating or Setting for Individual Motor-Compressor. The motor-compressor branch-circuit short-circuit and ground-fault protective device shall be capable of carrying the starting current of the motor. A protective device having a rating or setting not exceeding 175 percent of the motor-compressor rated-load current or branch-circuit selection current, whichever is greater, shall be permitted, provided that, where the protection specified is not sufficient for the starting current of the motor, the rating or setting shall be permitted to be increased but shall not exceed 225 percent of the motor rated-load current or branch-circuit selection current, whichever is greater.

(C) Protective Device Rating Not to Exceed the Manufacturer's Values. Where maximum protective device ratings shown on a manufacturer's overload relay table for use with a motor controller are less than the rating or setting selected in accordance with 440.22(A) and (B), the protective device rating shall not exceed the manufacturer's values marked on the equipment.

IV. BRANCH-CIRCUIT CONDUCTORS

440.31 General. The provisions of Part IV and Article 310 specify ampacities of conductors required to carry the motor current without overheating under the conditions specified, except as modified in 440.6(A), Exception No. 1.

 The provisions of these articles shall not apply to integral conductors of motors, to motor controllers and the like, or to conductors that form an integral part of approved equipment.

440.32 Single Motor-Compressor. Branch-circuit conductors supplying a single motor-compressor shall have an ampacity not less than 125 percent of either the motor-compressor rated-load current or the branch-circuit selection current, whichever is greater.

440.33 Motor-Compressor(s) With or Without Additional Motor Loads. Conductors supplying one or more motor-compressor(s) with or without an additional load(s) shall have an ampacity not less than the sum of the rated-load or branch-circuit selection current ratings, whichever is larger, of all the motor-compressors plus the full-load currents of the other motors, plus 25 percent of the highest motor or motor-compressor rating in the group.

Exception No. 1: Where the circuitry is interlocked so as to prevent the starting and running of a second motor-compressor or group of motor-compressors, the conductor size shall be determined from the largest motor-compressor or group of motor-compressors that is to be operated at a given time.

Exception No. 2: The branch-circuit conductors for room air conditioners shall be in accordance with Part VII of Article 440.

440.34 Combination Load. Conductors supplying a motor-compressor load in addition to other load(s) as calculated from Article 220 and other applicable articles shall have an ampacity sufficient for the other load(s) plus the required ampacity for the motor-compressor load determined in accordance with 440.33 or, for a single motor-compressor, in accordance with 440.32.

VI. MOTOR-COMPRESSOR AND BRANCH-CIRCUIT OVERLOAD PROTECTION

440.54 Motor-Compressors and Equipment on 15- or 20-Ampere Branch Circuits—Not Cord-and-Attachment-Plug-Connected. Overload protection for motor-compressors and equipment used on 15- or 20-ampere 120-volt, or 15-ampere 208- or 240-volt single-phase branch circuits as permitted in Article 210 shall be permitted as indicated in 440.54(A) and 440.54(B).

(A) Overload Protection. The motor-compressor shall be provided with overload protection selected as specified in 440.52(A). Both the controller and motor overload protective device shall be identified for installation with the short-circuit and ground-fault protective device for the branch circuit to which the equipment is connected.

(B) Time Delay. The short-circuit and ground-fault protective device protecting the branch circuit shall have sufficient time delay to permit the motor-compressor and other motors to start and accelerate their loads.

440.55 Cord-and-Attachment-Plug-Connected Motor-Compressors and Equipment on 15- or 20-Ampere Branch Circuits. Overload protection for motor-compressors and equipment that are cord-and-attachment-plug-connected and used on 15- or 20-ampere 120-volt, or 15-ampere 208- or 240-volt, single-phase branch circuits as permitted in Article 210 shall be permitted as indicated in 440.55(A), (B), and (C).

(A) Overload Protection. The motor-compressor shall be provided with overload protection as specified in 440.52(A). Both the controller and the motor overload protective device shall be identified for installation with the short-circuit and ground-fault protective device for the branch circuit to which the equipment is connected.

(B) Attachment Plug and Receptacle or Cord Connector Rating. The rating of the attachment plug and receptacle or cord connector shall not exceed 20 amperes at 125 volts or 15 amperes at 250 volts.

(C) Time Delay. The short-circuit and ground-fault protective device protecting the branch circuit shall have sufficient time delay to permit the motor-compressor and other motors to start and accelerate their loads.

VII. PROVISIONS FOR ROOM AIR CONDITIONERS

440.60 General. The provisions of Part VII shall apply to electrically energized room air conditioners that control temperature and humidity. For the purpose of Part VII, a room air conditioner (with or without provisions for heating) shall be considered as an ac appliance of the air-cooled window, console, or in-wall type that is installed in the conditioned room and that incorporates a hermetic refrigerant motor-compressor(s). The provisions of Part VII cover equipment rated not over 250 volts, single phase, and such equipment shall be permitted to be cord-and-attachment-plug-connected.

A room air conditioner that is rated 3-phase or rated over 250 volts shall be directly connected to a wiring method recognized in Chapter 3, and provisions of Part VII shall not apply.

440 61 Grounding. The enclosures of room air conditioners shall be connected to the equipment grounding conductor in accordance with 250.110, 250.112, and 250.114.

440.62 Branch-Circuit Requirements.

(A) Room Air Conditioner as a Single Motor Unit. A room air conditioner shall be considered as a single motor unit in determining its branch-circuit requirements where all the following conditions are met:

(1) It is cord-and-attachment-plug-connected.
(2) Its rating is not more than 40 amperes and 250 volts, single phase.
(3) Total rated-load current is shown on the room air-conditioner nameplate rather than individual motor currents.
(4) The rating of the branch-circuit short-circuit and ground-fault protective device does not exceed the ampacity of the branch-circuit conductors or the rating of the receptacle, whichever is less.

(B) Where No Other Loads Are Supplied. The total marked rating of a cord-and-attachment-plug-connected room air conditioner shall not exceed 80 percent of the rating of a branch circuit where no other loads are supplied.

(C) Where Lighting Units or Other Appliances Are Also Supplied. The total marked rating of a cord-and-attachment-plug-connected room air conditioner shall not exceed 50 percent of the rating of a branch circuit where lighting outlets, other appliances, or general-use receptacles are also supplied. Where the circuitry is interlocked to prevent simultaneous operation of the room air conditioner and energization of other outlets on the same branch circuit, a cord-and-attachment-plug-connected room air conditioner shall not exceed 80 percent of the branch-circuit rating.

440.63 Disconnecting Means. An attachment plug and receptacle or cord connector shall be permitted to serve as the disconnecting means for a single-phase room air conditioner rated 250 volts or less if (1) the manual controls on the room air conditioner are readily accessible and located within 1.8 m (6 ft) of the floor, or (2) an approved manually operable disconnecting means is installed in a readily accessible location within sight from the room air conditioner.

440.64 Supply Cords. Where a flexible cord is used to supply a room air conditioner, the length of such cord shall not exceed 3.0 m (10 ft) for a nominal, 120-volt rating or 1.8 m (6 ft) for a nominal, 208- or 240-volt rating.

440.65 Leakage-Current Detector-Interrupter (LCDI) and Arc-Fault Circuit Interrupter (AFCI). Single-phase cord-

and-plug-connected room air conditioners shall be provided with factory-installed LCDI or AFCI protection. The LCDI or AFCI protection shall be an integral part of the attachment plug or be located in the power supply cord within 300 mm (12 in.) of the attachment plug.

CHAPTER 23

COMMUNICATIONS CIRCUITS

INTRODUCTION

Chapter 23 contains provisions from Article 800—Communications Circuits. This article covers communications circuits and equipment. As defined in 800.2, a communications circuit is the circuit that extends voice, audio, video, data, interactive services, telegraph (except radio), outside wiring for fire/burglar alarms from the communications utility to the customer's communications equipment up to and including terminal equipment such as a telephone, fax machine, or answering machine.

This chapter contains four of the six parts in Article 800. Part I covers general installation requirements. Grounding methods are covered in Part IV, and Part V covers installation methods within buildings. The final part, Part VI, covers listing requirements for the different cable types.

ARTICLE 800
Communications Circuits

I. GENERAL

Informational Note: The general term grounding conductor as previously used in this article is replaced by either the term bonding conductor or the term grounding electrode conductor (GEC), where applicable, to more accurately reflect the application and function of the conductor. See Informational Note Figure 800(a) and Figure 800(b).

800.1 Scope. This article covers communications circuits and equipment.

800.18 Installation of Equipment. Equipment electrically connected to a communications network shall be listed in accordance with 800.170. Installation of equipment shall also comply with 110.3(B).

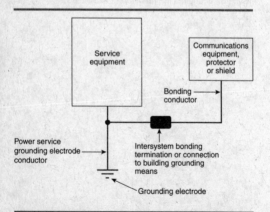

Informational Note Figure 800(a) Example of the Use of the Term *Bonding Conductor* Used in a Communications Installation.

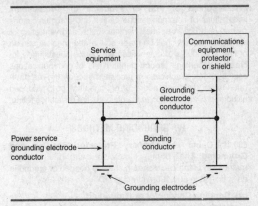

**Informational Note Figure 800(b) Example of the Use
of the Term *Grounding Electrode Conductor*
Used in a Communications Installation.**

**800.21 Access to Electrical Equipment Behind Panels
Designed to Allow Access.** Access to electrical equipment
shall not be denied by an accumulation of communications
wires and cables that prevents removal of panels, including
suspended ceiling panels.

800.24 Mechanical Execution of Work. Communications
circuits and equipment shall be installed in a neat and workmanlike manner. Cables installed exposed on the surface of
ceilings and sidewalls shall be supported by the building structure in such a manner that the cable will not be damaged by
normal building use. Such cables shall be secured by hardware, including straps, staples, cable ties, hangers, or similar
fittings designed and installed so as not to damage the cable.
The installation shall also conform to 300.4(D) and 300.11.

800.25 Abandoned Cables. The accessible portion of
abandoned communications cables shall be removed. Where
cables are identified for future use with a tag, the tag shall be
of sufficient durability to withstand the environment involved.

800.26 Spread of Fire or Products of Combustion.
Installations of communications cables and communications
raceways in hollow spaces, vertical shafts, and ventilation or
air-handling ducts shall be made so that the possible spread
of fire or products of combustion will not be substantially increased. Openings around penetrations of communications
cables and communications raceways through fire-resistant-
rated walls, partitions, floors, or ceilings shall be firestopped
using approved methods to maintain the fire resistance rating.

IV. GROUNDING METHODS

**800.100 Cable and Primary Protector Bonding and
Grounding.** The primary protector and the metallic
member(s) of the cable sheath shall be bonded or grounded
as specified in 800.100(A) through (D).

**(A) Bonding Conductor or Grounding
Electrode Conductor.**

(1) Insulation. The bonding conductor or grounding electrode conductor shall be listed and shall be permitted to be
insulated, covered, or bare.

(2) Material. The bonding conductor or grounding electrode
conductor shall be copper or other corrosion-resistant conductive material, stranded or solid.

(3) Size. The bonding conductor or grounding electrode
conductor shall not be smaller than 14 AWG. It shall have
a current-carrying capacity not less than the grounded metallic sheath member(s) and protected conductor(s) of the
communications cable. The bonding conductor or grounding
electrode conductor shall not be required to exceed 6 AWG.

(4) Length. The primary protector bonding conductor or
grounding electrode conductor shall be as short as practicable. In one- and two-family dwellings, the primary protector
bonding conductor or grounding electrode conductor shall be
as short as practicable, not to exceed 6.0 m (20 ft) in length.

*Exception: In one- and two-family dwellings where it is not
practicable to achieve an overall maximum primary protector bonding conductor or grounding electrode conductor*

length of 6.0 m (20 ft), a separate communications ground rod meeting the minimum dimensional criteria of 800.100(B) (3)(2) shall be driven, the primary protector shall be connected to the communications ground rod in accordance with 800.100(C), and the communications ground rod shall be connected to the power grounding electrode system in accordance with 800.100(D).

(5) Run in Straight Line. The bonding conductor or grounding electrode conductor shall be run in as straight a line as practicable.

(6) Physical Protection. Bonding conductors and grounding electrode conductors shall be protected where exposed to physical damage. Where the bonding conductor or grounding electrode conductor is installed in a metal raceway, both ends of the raceway shall be bonded to the contained conductor or to the same terminal or electrode to which the bonding conductor or grounding electrode conductor is connected.

(B) Electrode. The bonding conductor or grounding electrode conductor shall be connected in accordance with 800.100(B)(1), (B)(2), or (B)(3).

(1) In Buildings or Structures with an Intersystem Bonding Termination. If the building or structure served has an intersystem bonding termination as required by 250.94, the bonding conductor shall be connected to the intersystem bonding termination.

(2) In Buildings or Structures with Grounding Means. If the building or structure served has no intersystem bonding termination, the bonding conductor or grounding electrode conductor shall be connected to the nearest accessible location on one of the following:

(1) The building or structure grounding electrode system as covered in 250.50
(2) The grounded interior metal water piping system, within 1.5 m (5 ft) from its point of entrance to the building, as covered in 250.52
(3) The power service accessible means external to enclosures as covered in 250.94 Exception
(4) The nonflexible metallic power service raceway

(5) The service equipment enclosure

(6) The grounding electrode conductor or the grounding electrode conductor metal enclosure of the power service, or

(7) The grounding electrode conductor or the grounding electrode of a building or structure disconnecting means that is grounded to an electrode as covered in 250.32

A bonding device intended to provide a termination point for the grounding electrode conductor (intersystem bonding) shall not interfere with the opening of an equipment enclosure. A bonding device shall be mounted on non-removable parts. A bonding device shall not be mounted on a door or cover even if the door or cover is nonremovable.

For purposes of this section, the mobile home service equipment or the mobile home disconnecting means, as described in 800.90(B), shall be considered accessible.

(3) In Buildings or Structures Without an Intersystem Bonding Termination or Grounding Means. If the building or structure served has no intersystem bonding termination or grounding means, as described in 800.100(B)(2), the grounding electrode conductor shall be connected to either of the following:

(1) To any one of the individual electrodes described in 250.52(A)(1), (A)(2), (A)(3), or (A)(4)

(2) If the building or structure served has no intersystem bonding termination or has no grounding means, as described in 800.100(B)(2) or (B)(3)(1), to any one of the individual grounding electrodes described in 250.52(A)(7) and (A)(8) or to a ground rod or pipe not less than 1.5 m (5 ft) in length and 12.7 mm (1/2 in.) in diameter, driven, where practicable, into permanently damp earth and separated from lightning conductors as covered in 800.53 and at least 1.8 m (6 ft) from electrodes of other systems. Steam or hot water pipes or air terminal conductors (lightning-rod conductors) shall not be employed as electrodes for protectors and grounded metallic members.

(C) Electrode Connection. Connections to grounding electrodes shall comply with 250.70.

(D) Bonding of Electrodes. A bonding jumper not smaller than 6 AWG copper or equivalent shall be connected between the communications grounding electrode and power grounding electrode system at the building or structure served where separate electrodes are used.

Exception: At mobile homes as covered in 800.106.

V. INSTALLATION METHODS WITHIN BUILDINGS

800.110 Raceways for Communications Wires and Cables.

(A) Types of Raceways. Communications wires and cables shall be permitted to be installed in any raceway that complies with either (A)(1) or (A)(2).

(1) Raceways Recognized in Chapter 3. Communications wires and cables shall be permitted to be installed in any raceway included in Chapter 3. The raceways shall be installed in accordance with the requirements of Chapter 3.

(2) Other Permitted Raceways. Communications wires and cables shall be permitted to be installed in listed plenum communications raceway, listed riser communications raceway, or listed general-purpose communications raceway selected in accordance with the provisions of 800.113, and installed in accordance with 362.24 through 362.56, where the requirements applicable to electrical nonmetallic tubing apply.

(B) Raceway Fill for Communications Wires and Cables. The raceway fill requirements of Chapters 3 and 9 shall not apply to communications wires and cables.

800.113 Installation of Communications Wires, Cables and Raceways. Installation of communications wires, cables, and raceways shall comply with 800.113(A) through (L). Installation of raceways shall also comply with 800.110.

(A) Listing. Communications wires, cables, and raceways installed in buildings shall be listed.

(G) Risers—One- and Two-Family Dwellings. The following cables and raceways shall be permitted in one- and two-family dwellings:

(1) Types CMP, CMR, CMG, and CM cables
(2) Type CMX cable less than 6 mm (0.25 in.) in diameter
(3) Plenum, riser, and general-purpose communications raceways
(4) Types CMP, CMR, CMG, and CM cables installed in:

    a. Plenum communications raceway
    b. Riser communications raceway
    c. General-purpose communications raceway
    d. Riser cable routing assembly
    e. General-purpose cable routing assembly

(K) Multifamily Dwellings. The following cables, raceways, and wiring assemblies shall be permitted to be installed in multifamily dwellings in locations other than the locations covered in 800.113(B) through (G):

(1) Types CMP, CMR, CMG, and CM cables
(2) Type CMX cable less than 6 mm (0.25 in.) in diameter in nonconcealed spaces
(3) Plenum, riser, and general-purpose communications raceways
(4) Communications wires and Types CMP, CMR, CMG, and CM cables installed in:

    a. Plenum communications raceway
    b. Riser communications raceway
    c. General-purpose communications raceway

(5) Types CMP, CMR, CMG, and CM cables installed in:

    a. Riser cable routing assembly
    b. General-purpose cable routing assembly

(6) Communications wires and Types CMP, CMR, CMG, CM, and CMX cables installed in a raceway of a type recognized in Chapter 3
(7) Type CMUC undercarpet communications wires and cables installed under carpet

(L) One- and Two-Family Dwellings. The following cables and raceways shall be permitted to be installed in one- and two-family dwellings in locations other than the locations covered in 800.113(B) through (F):

(1) Types CMP, CMR, CMG, and CM cables

(2) Type CMX cable less than 6 mm (0.25 in.) in diameter
(3) Plenum, riser, and general-purpose communications raceways
(4) Communications wires and Types CMP, CMR, CMG, and CM cables installed in:

    a. Plenum communications raceway
    b. Riser communications raceway
    c. General-purpose communications raceway

(5) Types CMP, CMR, CMG, and CM cables installed in:

    a. Riser cable routing assembly
    b. General-purpose cable routing assembly

(6) Communication wires and Types CMP, CMR, CMG, CM, and CMX cables installed in a raceway of a type recognized in Chapter 3
(7) Type CMUC undercarpet communications wires and cables installed under carpet
(8) Hybrid power and communications cable listed in accordance with 800.179(I)

800.133 Installation of Communications Wires, Cables, and Equipment. Communications wires and cables from the protector to the equipment or, where no protector is required, communications wires and cables attached to the outside or inside of the building shall comply with 800.133(A) through (C).

(A) Separation from Other Conductors.

(1) In Raceways, Cable Trays, Boxes, Cables, and Enclosures.

(a) *Optical Fiber and Communications Cables.* Communications cables shall be permitted in the same raceway, cable tray, enclosure, or cable routing assembly with cables of any of the following:

(1) Nonconductive and conductive optical fiber cables in compliance with Parts I and IV of Article 770
(2) Community antenna television and radio distribution systems in compliance with Parts I and IV of Article 820

(3) Low-power network-powered broadband communications circuits in compliance with Parts I and IV of Article 830

(b) *Other Circuits.* Communications cables shall be permitted in the same raceway, cable tray, or enclosure with cables of any of the following:

(1) Class 2 and Class 3 remote-control, signaling, and power-limited circuits in compliance with Parts I and III of Article 725

(2) Power-limited fire alarm systems in compliance with Parts I and III of Article 760

(c) *Class 2 and Class 3 Circuits.* Class 1 circuits shall not be run in the same cable with communications circuits. Class 2 and Class 3 circuit conductors shall be permitted in the same cable with communications circuits, in which case the Class 2 and Class 3 circuits shall be classified as communications circuits and shall meet the requirements of this article. The cables shall be listed as communications cables.

(d) *Electric Light, Power, Class 1, Non–Power-Limited Fire Alarm, and Medium-Power Network-Powered Broadband Communications Circuits in Raceways, Compartments, and Boxes.* Communications conductors shall not be placed in any raceway, compartment, outlet box, junction box, or similar fitting with conductors of electric light, power, Class 1, non–power-limited fire alarm, or medium-power network-powered broadband communications circuits.

Exception No. 1: Where all of the conductors of electric light, power, Class 1, non–power-limited fire alarm, and medium-power network-powered broadband communications circuits are separated from all of the conductors of communications circuits by a permanent barrier or listed divider.

Exception No. 2: Power conductors in outlet boxes, junction boxes, or similar fittings or compartments where such conductors are introduced solely for power supply to communications equipment. The power circuit conductors shall be routed within the enclosure to maintain a minimum of 6 mm (0.25 in.) separation from the communications circuit conductors.

(B) Support of Communications Wires and Cables.
Raceways shall be used for their intended purpose. Communications wires and cables shall not be strapped, taped, or attached by any means to the exterior of any raceway as a means of support.

800.154 Applications of Listed Communications Wires, Cables, and Raceways. Permitted and nonpermitted applications of listed communications wires, cables, and raceways shall be as indicated in Table 800.154(a). The permitted applications shall be subject to the installation requirements of 800.110 and 800.113. The substitutions for communications cables listed in Table 800.154(b) and illustrated in Figure 800.154 shall be permitted.

800.156 Dwelling Unit Communications Outlet. For new construction, a minimum of one communications outlet shall be installed within the dwelling in a readily accessible area and cabled to the service provider demarcation point.

VI. LISTING REQUIREMENTS

800.170 Equipment. Communications equipment shall be listed as being suitable for electrical connection to a communications network.

(A) Primary Protectors. The primary protector shall consist of an arrester connected between each line conductor and ground in an appropriate mounting. Primary protector terminals shall be marked to indicate line and ground as applicable.

(B) Secondary Protectors. The secondary protector shall be listed as suitable to provide means to safely limit currents to less than the current-carrying capacity of listed indoor communications wire and cable, listed telephone set line cords, and listed communications terminal equipment having ports for external wire line communications circuits. Any overvoltage protection, arresters, or grounding connection shall be connected on the equipment terminals side of the secondary protector current-limiting means.

800.179 Communications Wires and Cables. Communications wires and cables shall be listed in accordance with

Table 800.154(a) Applications of Listed Communications Wires, Cables, and Raceways in Buildings

Applications	Wire, Cable, and Raceway Type									
	CMP	CMR	CMG CM	CMX	CMUC	Hybrid power and communications cables	Communications wires	Plenum communications raceways	Riser communications raceways	General-purpose communications raceways
In fabricated ducts as described in 300.22(B)										
In fabricated ducts	Y*	N	N	N	N	N	N	N	N	N
In metal raceway that complies with 300.22(B)	Y*	Y*	Y*	Y*	N	N	Y*	Y*	Y*	Y*
In other spaces used for environmental air as described in 300.22(C)										
In other spaces used for environmental air	Y*	N	N	N	N	N	N	Y*	N	N
In metal raceway that complies with 300.22(C)	Y*	Y*	Y*	Y*	N	N	Y*	Y*	Y*	Y*
In plenum communications raceways	Y*	N	N	N	N	N	N	Y*	N	N
Supported by open metal cable trays	Y*	N	N	N	N	N	N	Y*	N	N
Supported by solid bottom metal cable trays with solid metal covers	Y*	Y*	Y*	Y*	N	N	N	Y*	Y*	Y*

In risers

In vertical runs	Y*	Y*	N	N	N	N	Y*	N
In metal raceways	Y*	Y*	N	N	N	Y*	Y*	Y*
In fireproof shafts	Y*	Y*	N	N	N	Y*	Y*	Y*
In plenum communications raceways	Y*	N	N	N	N			N
In riser communications raceways	Y*	Y*	N	N	N			N
In riser cable routing assemblies	Y*	Y*	N	N	N	N	N	N
In one- and two-family dwellings	Y*	Y*	N	Y*	N	N	Y*	Y*

Within buildings in other than air-handling spaces and risers

General	Y*	Y*	Y*	N	N	N	Y*	Y*
In one- and two-family dwellings	Y*	Y*	Y*	N	Y*	Y*	Y*	Y*
In multifamily dwellings	Y*	Y*	Y*	N	Y*	Y*	Y*	Y*
In nonconcealed spaces	Y*	Y*	N	N	N	Y*	Y*	Y*
Supported by cable trays	Y*	Y*	N	N	Y*	Y*	Y*	Y*
Under carpet	N	N	N	N	N	N	N	N
In distributing frames and cross-connect arrays	Y*	Y*	Y*	Y*	N	Y*	Y*	Y*

(continued)

Table 800.154(a) Applications of Listed Communications Wires, Cables, and Raceways in Buildings *Continued*

Applications	Wire, Cable, and Raceway Type									
	CMP	CMR	CMG CM	CMX	CMUC	Hybrid power and communications cables	Communications wires	Plenum communications raceways	Riser communications raceways	General-purpose communications raceways
In any raceway recognized in Chapter 3	Y*	Y*	Y*	Y*	N	N		Y*	Y*	Y*
In plenum communications raceways	Y*	Y*	Y*	N	N	N	Y*			
In riser communications raceways and riser cable routing assemblies	Y*	Y*	Y*	N	N	N	Y*			
In general-purpose communications raceways and general-purpose cable routing assemblies	Y*	Y*	Y*	N	N.	N	Y*			

Note: An "N" in the table indicates that the cable type is not permitted to be installed in the application. A "Y*" indicates that the cable is permitted to be installed in the application, subject to the limitations described in 800.110 and 800.113.

Informational Note 1: Part V of Article 800 covers installation methods within buildings. This table covers the applications of listed communications wires, cables and raceways in buildings. The definition of point of entrance is in 800.2. Communications entrance cables that have not emerged from the rigid metal conduit or intermediate metal conduit are not considered to be in the building.

Informational Note No. 2: For information on the restrictions to the installation of communications cables in fabricated ducts see 800.113(B).

Table 800.154(b) Cable Substitutions

Cable Type	Permitted Substitutions
CMR	CMP
CMG, CM	CMP, CMR
CMX	CMP, CMR, CMG, CM

800.179(A) through (I) and marked in accordance with Table 800.179. Conductors in communications cables, other than in a coaxial cable, shall be copper.

Communications wires and cables shall have a voltage rating of not less than 300 volts. The insulation for the individual conductors, other than the outer conductor of a coaxial cable, shall be rated for 300 volts minimum. The cable voltage rating shall not be marked on the cable or on the undercarpet communications wire. Communications wires and cables shall have a temperature rating of not less than 60°C.

Exception: Voltage markings shall be permitted where the cable has multiple listings and voltage marking is required for one or more of the listings.

(A) Type CMP. Type CMP communications plenum cables shall be listed as being suitable for use in ducts, plenums, and other spaces used for environmental air and shall also be listed as having adequate fire-resistant and low smoke-producing characteristics.

(B) Type CMR. Type CMR communications riser cables shall be listed as being suitable for use in a vertical run in a shaft or from floor to floor and shall also be listed as having

Table 800.179 Cable Markings

Cable Marking	Type
CMP	Communications plenum cable
CMR	Communications riser cable
CMG	Communications general-purpose cable
CM	Communications general-purpose cable
CMX	Communications cable, limited use
CMUC	Undercarpet communications wire and cable

fire-resistant characteristics capable of preventing the carrying of fire from floor to floor.

(C) Type CMG. Type CMG general-purpose communications cables shall be listed as being suitable for general-purpose communications use, with the exception of risers and plenums, and shall also be listed as being resistant to the spread of fire.

(D) Type CM. Type CM communications cables shall be listed as being suitable for general-purpose communications use, with the exception of risers and plenums, and shall also be listed as being resistant to the spread of fire.

(E) Type CMX. Type CMX limited-use communications cables shall be listed as being suitable for use in dwellings and for use in raceway and shall also be listed as being resistant to flame spread.

(F) Type CMUC Undercarpet Wires and Cables. Type CMUC undercarpet communications wires and cables shall be listed as being suitable for undercarpet use and shall also be listed as being resistant to flame spread.

(G) Communications Circuit Integrity (CI) Cables. Cables suitable for use in communications systems to ensure survivability of critical circuits during a specified time under fire conditions shall be listed as circuit integrity (CI) cable. Cables identified in 800.179(A) through (E) that meet the requirements for circuit integrity shall have the additional classification using the suffix "CI."

(H) Communications Wires. Communications wires, such as distributing frame wire and jumper wire, shall be listed as being resistant to the spread of fire.

(I) Hybrid Power and Communications Cables. Listed hybrid power and communications cables shall be permitted where the power cable is a listed Type NM or NM-B conforming to the provisions of Article 334, and the communications cable is a listed Type CM, the jackets on the listed NM or NM-B and listed CM cables are rated for 600 volts minimum, and the hybrid cable is listed as being resistant to the spread of fire.

800.182 Communications Raceways and Cable Routing Assemblies. Communications raceways shall be listed in accordance with 800.182(A) through (C).

(A) Plenum Communications Raceways. Plenum communications raceways listed as plenum optical fiber raceways shall be permitted for use in ducts, plenums, and other spaces used for environmental air and shall also be listed as having adequate fire-resistant and low smoke-producing characteristics.

(B) Riser Communications Raceways and Cable Routing Assemblies. Riser communications raceways and riser cable routing assemblies shall be listed as having adequate fire-resistant characteristics capable of preventing the carrying of fire from floor to floor.

(C) General-Purpose Communications Raceways and Cable Routing Assemblies. General-purpose communications raceways and cable routing assemblies shall be listed as being resistant to the spread of fire.

CHAPTER 24

SEPARATE BUILDINGS OR STRUCTURES

INTRODUCTION

This chapter contains provisions from two sections of Article 225—Outside Branch Circuits and Feeders and from one section of Article 250—Grounding and Bonding. This chapter covers the section pertaining to two or more buildings or structures supplied from a common service. This section's requirements cover: grounding electrode(s), grounded systems, disconnecting means located in a separate building or structure on the same premises, and the grounding electrode conductor.

ARTICLE 225

Outside Branch Circuits and Feeders

225.31 Disconnecting Means. Means shall be provided for disconnecting all ungrounded conductors that supply or pass through the building or structure.

225.32 Location. The disconnecting means shall be installed either inside or outside of the building or structure served or where the conductors pass through the building or structure. The disconnecting means shall be at a readily accessible location nearest the point of entrance of the conductors. For the purposes of this section, the requirements in 230.6 shall be utilized.

Exception No. 1: For installations under single management, where documented safe switching procedures are established and maintained for disconnection, and where the installation is monitored by qualified individuals, the disconnecting means shall be permitted to be located elsewhere on the premises.

Exception No. 3: For towers or poles used as lighting standards, the disconnecting means shall be permitted to be located elsewhere on the premises.

ARTICLE 250

Grounding and Bonding

250.32 Buildings or Structures Supplied by a Feeder(s) or Branch Circuit(s).

(A) Grounding Electrode. Building(s) or structure(s) supplied by feeder(s) or branch circuit(s) shall have a grounding electrode or grounding electrode system installed in accordance with Part III of Article 250. The grounding electrode conductor(s) shall be connected in accordance with 250.32(B) or (C). Where there is no existing grounding electrode, the grounding electrode(s) required in 250.50 shall be installed.

Exception: A grounding electrode shall not be required where only a single branch circuit, including a multiwire branch cir-

cuit, supplies the building or structure and the branch circuit includes an equipment grounding conductor for grounding the normally non–current-carrying metal parts of equipment.

(B) Grounded Systems.

(1) Supplied by a Feeder or Branch Circuit. An equipment grounding conductor, as described in 250.118, shall be run with the supply conductors and be connected to the building or structure disconnecting means and to the grounding electrode(s). The equipment grounding conductor shall be used for grounding or bonding of equipment, structures, or frames required to be grounded or bonded. The equipment grounding conductor shall be sized in accordance with 250.122. Any installed grounded conductor shall not be connected to the equipment grounding conductor or to the grounding electrode(s).

Exception: For installations made in compliance with previous editions of this Code that permitted such connection, the grounded conductor run with the supply to the building or structure shall be permitted to serve as the ground-fault return path if all of the following requirements continue to be met:

(1) *An equipment grounding conductor is not run with the supply to the building or structure.*
(2) *There are no continuous metallic paths bonded to the grounding system in each building or structure involved.*
(3) *Ground-fault protection of equipment has not been installed on the supply side of the feeder(s).*

If the grounded conductor is used for grounding in accordance with the provision of this exception, the size of the grounded conductor shall not be smaller than the larger of either of the following:

(1) *That required by 220.61*
(2) *That required by 250.122*

(2) Supplied by Separately Derived System.

(a) *With Overcurrent Protection.* If overcurrent protection is provided where the conductors originate, the installation shall comply with 250.32(B)(1).

(b) *Without Overcurrent Protection.* If overcurrent protection is not provided where the conductors originate, the installation shall comply with 250.30(A). If installed, the supply-side bonding jumper shall be connected to the building or structure disconnecting means and to the grounding electrode(s).

(D) Disconnecting Means Located in Separate Building or Structure on the Same Premises. Where one or more disconnecting means supply one or more additional buildings or structures under single management, and where these disconnecting means are located remote from those buildings or structures in accordance with the provisions of 225.32, Exception No. 1 and No. 2, 700.12(B)(6), 701.12(B)(5), or 702.12, all of the following conditions shall be met:

(1) The connection of the grounded conductor to the grounding electrode, to normally non–current-carrying metal parts of equipment, or to the equipment grounding conductor at a separate building or structure shall not be made.

(2) An equipment grounding conductor for grounding and bonding any normally non–current-carrying metal parts of equipment, interior metal piping systems, and building or structural metal frames is run with the circuit conductors to a separate building or structure and connected to existing grounding electrode(s) required in Part III of this article, or, where there are no existing electrodes, the grounding electrode(s) required in Part III of this article shall be installed where a separate building or structure is supplied by more than one branch circuit.

(3) The connection between the equipment grounding conductor and the grounding electrode at a separate building or structure shall be made in a junction box, panelboard, or similar enclosure located immediately inside or outside the separate building or structure.

(E) Grounding Electrode Conductor. The size of the grounding electrode conductor to the grounding electrode(s) shall not be smaller than given in 250.66, based on the largest ungrounded supply conductor. The installation shall comply with Part III of this article.

CHAPTER 25

SWIMMING POOLS, FOUNTAINS, AND SIMILAR INSTALLATIONS

INTRODUCTION

This chapter contains provisions from Article 680—Swimming Pools, Fountains, and Similar Installations. The provisions of this article apply to the construction and installation of electric wiring for and equipment in (or adjacent to) all swimming, wading, and decorative pools, fountains, hot tubs, spas, and hydromassage bathtubs, whether permanently installed or storable, as well as to metallic auxiliary equipment (such as pumps, filters, etc.). The requirements of Article 680 pertain to grounding, bonding, conductor clearances, receptacles, lighting, equipment, and ground-fault circuit-interrupter protection for personnel.

Contained in this pocket guide are six of the seven parts that are included in Article 680. The six parts are: I General; II Permanently Installed Pools; III Storable Pools; IV Spas and Hot Tubs; V Fountains; and VII Hydro-massage Bathtubs.

ARTICLE 680
Swimming Pools, Fountains, and Similar Installations

I. GENERAL

680.1 Scope. The provisions of this article apply to the construction and installation of electrical wiring for, and equipment in or adjacent to, all swimming, wading, therapeutic, and decorative pools; fountains; hot tubs; spas; and hydromassage bathtubs, whether permanently installed or storable, and to metallic auxiliary equipment, such as pumps, filters, and similar equipment. The term *body of water* used throughout Part I applies to all bodies of water covered in this scope unless otherwise amended.

680.6 Grounding. Electrical equipment shall be grounded in accordance with Parts V, VI, and VII of Article 250 and connected by wiring methods of Chapter 3, except as modified by this article. The following equipment shall be grounded:

(1) Through-wall lighting assemblies and underwater luminaires, other than those low-voltage lighting products listed for the application without a grounding conductor
(2) All electrical equipment located within 1.5 m (5 ft) of the inside wall of the specified body of water
(3) All electrical equipment associated with the recirculating system of the specified body of water
(4) Junction boxes
(5) Transformer and power supply enclosures
(6) Ground-fault circuit interrupters
(7) Panelboards that are not part of the service equipment and that supply any electrical equipment associated with the specified body of water

680.8 Overhead Conductor Clearances. Overhead conductors shall meet the clearance requirements in this section. Where a minimum clearance from the water level is given, the measurement shall be taken from the maximum water level of the specified body of water.

(A) Power. With respect to service drop conductors and open overhead wiring, swimming pool and similar instal-

Table 680.8 Overhead Conductor Clearances

Clearance Parameters	Insulated Cables, 0–750 Volts to Ground, Supported on and Cabled Together with a Solidly Grounded Bare Messenger or Solidly Grounded Neutral Conductor		All Other Conductors Voltage to Ground			
			0 through 15 kV		Over 15 through 50 kV	
	m	ft	m	ft	m	ft
A. Clearance in any direction to the water level, edge of water surface, base of diving platform, or permanently anchored raft	6.9	22.5	7.5	25	8.0	27
B. Clearance in any direction to the observation stand, tower, or diving platform	4.4	14.5	5.2	17	5.5	18
C. Horizontal limit of clearance measured from inside wall of the pool	This limit shall extend to the outer edge of the structures listed in A and B of this table but not to less than 3 m (10 ft).					

lations shall comply with the minimum clearances given in Table 680.8 and illustrated in Figure 680.8.

(B) Communications Systems. Communications, radio, and television coaxial cables within the scope of Articles 800 through 820 shall be permitted at a height of not less than 3.0 m (10 ft) above swimming and wading pools, diving structures, and observation stands, towers, or platforms.

(C) Network-Powered Broadband Communications Systems. The minimum clearances for overhead network-powered broadband communications systems conductors from pools or fountains shall comply with the provisions in Table 680.8 for conductors operating at 0 to 750 volts to ground.

680.10 Underground Wiring Location. Underground wiring shall not be permitted under the pool or within the area extending 1.5 m (5 ft) horizontally from the inside wall of the pool unless this wiring is necessary to supply pool equipment permitted by this article. Where space limitations prevent wiring from being routed a distance 1.5 m (5 ft) or more from the pool, such wiring shall be permitted where installed in complete raceway systems of rigid metal conduit, intermediate metal conduit, or a nonmetallic raceway system. All metal conduit shall be corrosion resistant and suitable for the location. The minimum cover depth shall be as given in Table 680.10.

680.12 Maintenance Disconnecting Means. One or more means to simultaneously disconnect all ungrounded conduc-

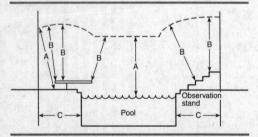

Figure 680.8 Clearances from Pool Structures.

tors shall be provided for all utilization equipment other than lighting. Each means shall be readily accessible and within sight from its equipment and shall be located at least 1.5 m (5 ft) horizontally from the inside walls of a pool, spa, or hot tub unless separated from the open water by a permanently installed barrier that provides a 1.5 m (5 ft) reach path or greater. This horizontal distance is to be measured from the water's edge along the shortest path required to reach the disconnect.

II. PERMANENTLY INSTALLED POOLS

680.20 General. Electrical installations at permanently installed pools shall comply with the provisions of Part I and Part II of this article.

680.21 Motors.

(A) Wiring Methods. The wiring to a pool motor shall comply with (A)(1) unless modified for specific circumstances by (A)(2), (A)(3), (A)(4), or (A)(5).

(1) General. The branch circuits for pool-associated motors shall be installed in rigid metal conduit, intermediate metal conduit, rigid polyvinyl chloride conduit, reinforced thermosetting resin conduit, or Type MC cable listed for the location. Other wiring methods and materials shall be permitted in specific locations or applications as covered in this section. Any wiring method employed shall contain an insulated copper equipment grounding conductor sized in accordance with 250.122 but not smaller than 12 AWG.

(2) On or Within Buildings. Where installed on or within buildings, electrical metallic tubing shall be permitted.

(3) Flexible Connections. Where necessary to employ flexible connections at or adjacent to the motor, liquidtight flexible metal or liquidtight flexible nonmetallic conduit with approved fittings shall be permitted.

(4) One-Family Dwellings. In the interior of dwelling units, or in the interior of accessory buildings associated with a dwelling unit, any of the wiring methods recognized in Chapter 3 of this *Code* that comply with the provisions of this section shall be permitted. Where run in a cable assembly, the equipment grounding conductor shall be permitted to be uninsulated, but it shall be enclosed within the outer sheath of the cable assembly.

(5) Cord-and-Plug Connections. Pool-associated motors shall be permitted to employ cord-and-plug connections. The flexible cord shall not exceed 900 mm (3 ft) in length. The flexible cord shall include a copper equipment grounding conductor sized in accordance with 250.122 but not smaller than 12 AWG. The cord shall terminate in a grounding-type attachment plug.

(B) Double Insulated Pool Pumps. A listed cord-and-plug-connected pool pump incorporating an approved system of double insulation that provides a means for grounding only the internal and nonaccessible, non–current-carrying metal parts of the pump shall be connected to any wiring method recognized in Chapter 3 that is suitable for the location. Where the bonding grid is connected to the equipment grounding conductor of the motor circuit in accordance with the second sentence of 680.26(B)(6)(a), the branch-circuit wiring shall comply with 680.21(A).

(C) GFCI Protection. Outlets supplying pool pump motors connected to single-phase, 120 volt through 240 volt branch circuits, rated 15 or 20 amperes, whether by receptacle or by direct connection, shall be provided with ground-fault circuit-interrupter protection for personnel.

680.22 Lighting, Receptacles, and Equipment.

(A) Receptacles.

(1) Circulation and Sanitation System, Location. Receptacles that provide power for water-pump motors or for other loads directly related to the circulation and sanitation system shall be located at least 3.0 m (10 ft) from the inside walls of the pool, or not less than 1.83 m (6 ft) from the inside walls of the pool if they meet all of the following conditions:

(1) Consist of single receptacles
(2) Employ a locking configuration
(3) Are of the grounding type
(4) Have GFCI protection

(2) Other Receptacles, Location. Other receptacles shall be not less than 1.83 m (6 ft) from the inside walls of a pool.

(3) Dwelling Unit(s). Where a permanently installed pool is installed at a dwelling unit(s), no fewer than one 125-volt, 15- or 20-ampere receptacle on a general-purpose branch circuit shall be located not less than 1.83 m (6 ft) from, and not more than 6.0 m (20 ft) from, the inside wall of the pool. This receptacle shall be located not more than 2.0 m (6 ft 6 in.) above the floor, platform, or grade level serving the pool.

(4) GFCI Protection. All 15- and 20-ampere, single-phase, 125-volt receptacles located within 6.0 m (20 ft) of the inside walls of a pool shall be protected by a ground-fault circuit interrupter.

(5) Measurements. In determining the dimensions in this section addressing receptacle spacings, the distance to be measured shall be the shortest path the supply cord of an appliance connected to the receptacle would follow without piercing a floor, wall, ceiling, doorway with hinged or sliding door, window opening, or other effective permanent barrier.

(B) Luminaires, Lighting Outlets, and Ceiling-Suspended (Paddle) Fans.

(1) New Outdoor Installation Clearances. In outdoor pool areas, luminaires, lighting outlets, and ceiling-suspended (paddle) fans installed above the pool or the area extending 1.5 m (5 ft) horizontally from the inside walls of the pool shall be installed at a height not less than 3.7 m (12 ft) above the maximum water level of the pool.

(2) Indoor Clearances. For installations in indoor pool areas, the clearances shall be the same as for outdoor areas unless modified as provided in this paragraph. If the branch circuit supplying the equipment is protected by a ground-fault circuit interrupter, the following equipment shall be permitted at a height not less than 2.3 m (7 ft 6 in.) above the maximum pool water level:

(1) Totally enclosed luminaires
(2) Ceiling-suspended (paddle) fans identified for use beneath ceiling structures such as provided on porches or patios

(3) Existing Installations. Existing luminaires and lighting outlets located less than 1.5 m (5 ft) measured horizontally from the inside walls of a pool shall be not less than 1.5 m (5 ft) above the surface of the maximum water level, shall be rigidly attached to the existing structure, and shall be protected by a ground-fault circuit interrupter.

(4) GFCI Protection in Adjacent Areas. Luminaires, lighting outlets, and ceiling-suspended (paddle) fans installed in the area extending between 1.5 m (5 ft) and 3.0 m (10 ft) horizontally from the inside walls of a pool shall be protected by a ground-fault circuit interrupter unless installed not less than 1.5 m (5 ft) above the maximum water level and rigidly attached to the structure adjacent to or enclosing the pool.

(5) Cord-and-Plug-Connected Luminaires. Cord-and-plug-connected luminaires shall comply with the requirements of 680.7 where installed within 4.9 m (16 ft) of any point on the water surface, measured radially.

(C) Switching Devices. Switching devices shall be located at least 1.5 m (5 ft) horizontally from the inside walls of a pool unless separated from the pool by a solid fence, wall, or other permanent barrier. Alternatively, a switch that is listed as being acceptable for use within 1.5 m (5 ft) shall be permitted.

(D) Other Outlets. Other outlets shall be not less than 3.0 m (10 ft) from the inside walls of the pool. Measurements shall be determined in accordance with 680.22(A)(5).

680.23 Underwater Luminaires. This section covers all luminaires installed below the normal water level of the pool.

(A) General.

(1) Luminaire Design, Normal Operation. The design of an underwater luminaire supplied from a branch circuit either directly or by way of a transformer or power supply meeting the requirements of this section shall be such that, where the luminaire is properly installed without a ground-fault circuit interrupter, there is no shock hazard with any likely combination of fault conditions during normal use (not relamping).

(2) Transformers and Power Supplies. Transformers and power supplies used for the supply of underwater luminaires, together with the transformer or power supply enclosure, shall be listed for swimming pool and spa use. The transformer or power supply shall incorporate either a transformer of the isolated winding type, with an ungrounded secondary that has a grounded metal barrier between the primary and secondary windings, or one that incorporates an approved system of double insulation between the primary and secondary windings.

(3) GFCI Protection, Relamping. A ground-fault circuit interrupter shall be installed in the branch circuit supplying luminaires operating at more than the low voltage contact limit such that there is no shock hazard during relamping. The installation of the ground-fault circuit interrupter shall be such that there is no shock hazard with any likely fault-condition combination that involves a person in a conductive path from any ungrounded part of the branch circuit or the luminaire to ground.

(4) Voltage Limitation. No luminaires shall be installed for operation on supply circuits over 150 volts between conductors.

(5) Location, Wall-Mounted Luminaires. Luminaires mounted in walls shall be installed with the top of the luminaire lens not less than 450 mm (18 in.) below the normal water level of the pool, unless the luminaire is listed and identified for use at lesser depths. No luminaire shall be installed less than 100 mm (4 in.) below the normal water level of the pool.

(6) Bottom-Mounted Luminaires. A luminaire facing upward shall comply with either (1) or (2):

(1) Have the lens adequately guarded to prevent contact by any person
(2) Be listed for use without a guard

(7) Dependence on Submersion. Luminaires that depend on submersion for safe operation shall be inherently protected against the hazards of overheating when not submerged.

(8) Compliance. Compliance with these requirements shall be obtained by the use of a listed underwater luminaire and by installation of a listed ground-fault circuit interrupter in the branch circuit or a listed transformer or power supply for luminaires operating at not more than the low voltage contact limit.

(B) Wet-Niche Luminaires.

(1) Forming Shells. Forming shells shall be installed for the mounting of all wet-niche underwater luminaires and shall be equipped with provisions for conduit entries. Metal parts of the luminaire and forming shell in contact with the pool water shall be of brass or other approved corrosion-resistant metal. All forming shells used with nonmetallic conduit systems, other than those that are part of a listed low-voltage lighting system not requiring grounding, shall include provisions for terminating an 8 AWG copper conductor.

(2) Wiring Extending Directly to the Forming Shell. Conduit shall be installed from the forming shell to a junction box or other enclosure conforming to the requirements in 680.24. Conduit shall be rigid metal, intermediate metal, liquidtight flexible nonmetallic, or rigid nonmetallic.

(a) *Metal Conduit.* Metal conduit shall be approved and shall be of brass or other approved corrosion-resistant metal.

(b) *Nonmetallic Conduit.* Where a nonmetallic conduit is used, an 8 AWG insulated solid or stranded copper bonding jumper shall be installed in this conduit unless a listed low-voltage lighting system not requiring grounding is used. The bonding jumper shall be terminated in the forming shell, junction box or transformer enclosure, or ground-fault circuit-interrupter enclosure. The termination of the 8 AWG bonding jumper in the forming shell shall be covered with, or encapsulated in, a listed potting compound to protect the connection from the possible deteriorating effect of pool water.

(3) Equipment Grounding Provisions for Cords. Other than listed low-voltages lighting systems not requiring grounding wet-niche luminaires that are supplied by a flexible cord or cable shall have all exposed non–current-carrying metal parts grounded by an insulated copper equipment grounding conductor that is an integral part of the cord or cable. This grounding conductor shall be connected to a grounding terminal in the supply junction box, transformer enclosure, or other enclosure. The grounding conductor shall not be smaller than the supply conductors and not smaller than 16 AWG.

(4) Luminaire Grounding Terminations. The end of the flexible-cord jacket and the flexible-cord conductor terminations within a luminaire shall be covered with, or encapsulated in, a suitable potting compound to prevent the entry of water into the luminaire through the cord or its conductors. If present, the grounding connection within a luminaire shall be similarly treated to protect such connection from the deteriorating effect of pool water in the event of water entry into the luminaire.

(5) Luminaire Bonding. The luminaire shall be bonded to, and secured to, the forming shell by a positive locking device that ensures a low-resistance contact and requires a tool to remove the luminaire from the forming shell. Bonding shall not be required for luminaires that are listed for the application and have no non–current-carrying metal parts.

(6) Servicing. All wet-niche luminaires shall be removable from the water for inspection, relamping, or other maintenance. The forming shell location and length of cord in the forming shell shall permit personnel to place the removed luminaire on the deck or other dry location for such maintenance. The luminaire maintenance location shall be accessible without entering or going in the pool water.

(C) Dry-Niche Luminaires.

(1) Construction. A dry-niche luminaire shall have provision for drainage of water. Other than listed low voltage luminaires not requiring grounding, a dry-niche luminaire shall have means for accommodating one equipment grounding conductor for each conduit entry.

(2) Junction Box. A junction box shall not be required but, if used, shall not be required to be elevated or located as specified in 680.24(A)(2) if the luminaire is specifically identified for the purpose.

(D) No-Niche Luminaires. A no-niche luminaire shall meet the construction requirements of 680.23(B)(3) and be installed in accordance with the requirements of 680.23(B). Where connection to a forming shell is specified, the connection shall be to the mounting bracket.

(E) Through-Wall Lighting Assembly. A through-wall lighting assembly shall be equipped with a threaded entry or hub, or a nonmetallic hub, for the purpose of accommodating the termination of the supply conduit. A through-wall lighting assembly shall meet the construction requirements of 680.23(B)(3) and be installed in accordance with the requirements of 680.23. Where connection to a forming shell is specified, the connection shall be to the conduit termination point.

(F) Branch-Circuit Wiring.

(1) Wiring Methods. Branch-circuit wiring on the supply side of enclosures and junction boxes connected to conduits run to wet-niche and no-niche luminaires, and the field wiring compartments of dry-niche luminaires, shall be installed using rigid metal conduit, intermediate metal conduit, liquidtight flexible nonmetallic conduit, rigid polyvinyl chloride conduit, or reinforced thermosetting resin conduit. Where installed on buildings, electrical metallic tubing shall be permitted, and where installed within buildings, electrical nonmetallic tubing, Type MC cable, electrical metallic tubing, or Type AC cable shall be permitted. In all cases, an insulated equipment grounding conductor sized in accordance with Table 250.122 but not less than 12 AWG shall be required.

Exception: Where connecting to transformers for pool lights, liquidtight flexible metal conduit shall be permitted. The length shall not exceed 1.8 m (6 ft) for any one length or exceed 3.0 m (10 ft) in total length used.

(2) Equipment Grounding. Other than listed low-voltage luminaires not requiring grounding, all through-wall lighting assemblies, wet-niche, dry-niche, or no-niche luminaires shall be connected to an insulated copper equipment grounding conductor installed with the circuit conductors. The equipment grounding conductor shall be installed without joint or splice except as permitted in (F)(2)(a) and (F)(2)(b). The equipment grounding conductor shall be sized in accordance with Table 250.122 but shall not be smaller than 12 AWG.

Exception: An equipment grounding conductor between the wiring chamber of the secondary winding of a transformer and a junction box shall be sized in accordance with the overcurrent device in this circuit.

(a) If more than one underwater luminaire is supplied by the same branch circuit, the equipment grounding conductor, installed between the junction boxes, transformer enclosures, or other enclosures in the supply circuit to wet-niche luminaires, or between the field-wiring compartments of dry-niche luminaires, shall be permitted to be terminated on grounding terminals.

(b) If the underwater luminaire is supplied from a transformer, ground-fault circuit interrupter, clock-operated switch, or a manual snap switch that is located between the panelboard and a junction box connected to the conduit that extends directly to the underwater luminaire, the equipment grounding conductor shall be permitted to terminate on grounding terminals on the transformer, ground-fault circuit interrupter, clock-operated switch enclosure, or an outlet box used to enclose a snap switch.

(3) Conductors. Conductors on the load side of a ground-fault circuit interrupter or of a transformer, used to comply with the provisions of 680.23(A)(8), shall not occupy raceways, boxes, or enclosures containing other conductors unless one of the following conditions applies:

(1) The other conductors are protected by ground-fault circuit interrupters.
(2) The other conductors are grounding conductors.
(3) The other conductors are supply conductors to a feed-through-type ground-fault circuit interrupter.
(4) Ground-fault circuit interrupters shall be permitted in a panelboard that contains circuits protected by other than ground-fault circuit interrupters.

680.24 Junction Boxes and Electrical Enclosures for Transformers or Ground-Fault Circuit Interrupters.

(A) Junction Boxes. A junction box connected to a conduit that extends directly to a forming shell or mounting bracket of a no-niche luminaire shall meet the requirements of this section.

(1) Construction. The junction box shall be listed as a swimming pool junction box and shall comply with the following conditions:

(1) Be equipped with threaded entries or hubs or a nonmetallic hub
(2) Be comprised of copper, brass, suitable plastic, or other approved corrosion-resistant material
(3) Be provided with electrical continuity between every connected metal conduit and the grounding terminals by means of copper, brass, or other approved corrosion-resistant metal that is integral with the box

(2) Installation. Where the luminaire operates over the low voltage contact limit, the junction box location shall comply with (A)(2)(a) and (A)(2)(b). Where the luminaire operates at the low voltage contact limit or less, the junction box location shall be permitted to comply with (A)(2)(c).

(a) *Vertical Spacing.* The junction box shall be located not less than 100 mm (4 in.), measured from the inside of the bottom of the box, above the ground level, or pool deck, or not less than 200 mm (8 in.) above the maximum pool water level, whichever provides the greater elevation.

(b) *Horizontal Spacing.* The junction box shall be located not less than 1.2 m (4 ft) from the inside wall of the pool, unless separated from the pool by a solid fence, wall, or other permanent barrier.

(c) *Flush Deck Box.* If used on a lighting system operating at the low voltage contact limit or less, a flush deck box shall be permitted if both of the following conditions are met:

(1) An approved potting compound is used to fill the box to prevent the entrance of moisture.
(2) The flush deck box is located not less than 1.2 m (4 ft) from the inside wall of the pool.

(B) Other Enclosures. An enclosure for a transformer, ground-fault circuit interrupter, or a similar device connected to a conduit that extends directly to a forming shell or mounting bracket of a no-niche luminaire shall meet the requirements of this section.

(1) Construction. The enclosure shall be listed and labeled for the purpose and meet the following requirements:

(1) Equipped with threaded entries or hubs or a nonmetallic hub
(2) Comprised of copper, brass, suitable plastic, or other approved corrosion-resistant material
(3) Provided with an approved seal, such as duct seal at the conduit connection, that prevents circulation of air between the conduit and the enclosures
(4) Provided with electrical continuity between every connected metal conduit and the grounding terminals by means of copper, brass, or other approved corrosion-resistant metal that is integral with the box

(2) Installation.

(a) *Vertical Spacing.* The enclosure shall be located not less than 100 mm (4 in.), measured from the inside of the bottom of the box, above the ground level, or pool deck, or not less than 200 mm (8 in.) above the maximum pool water level, whichever provides the greater elevation.

(b) *Horizontal Spacing.* The enclosure shall be located not less than 1.2 m (4 ft) from the inside wall of the pool, unless separated from the pool by a solid fence, wall, or other permanent barrier.

(C) Protection. Junction boxes and enclosures mounted above the grade of the finished walkway around the pool shall not be located in the walkway unless afforded additional protection, such as by location under diving boards, adjacent to fixed structures, and the like.

(D) Grounding Terminals. Junction boxes, transformer and power-supply enclosures, and ground-fault circuit-interrupter enclosures connected to a conduit that extends directly to a forming shell or mounting bracket of a no-niche luminaire shall be provided with a number of grounding terminals that shall be no fewer than one more than the number of conduit entries.

(E) Strain Relief. The termination of a flexible cord of an underwater luminaire within a junction box, transformer or power-supply enclosure, ground-fault circuit interrupter, or other enclosure shall be provided with a strain relief.

(F) Grounding. The equipment grounding conductor terminals of a junction box, transformer enclosure, or other enclosure in the supply circuit to a wet-niche or no-niche luminaire and the field-wiring chamber of a dry-niche luminaire shall be connected to the equipment grounding terminal of the panelboard. This terminal shall be directly connected to the panelboard enclosure.

680.25 Feeders. These provisions shall apply to any feeder on the supply side of panelboards supplying branch circuits for pool equipment covered in Part II of this article and on the load side of the service equipment or the source of a separately derived system.

(A) Wiring Methods.

(1) Feeders. Feeders shall be installed in rigid metal conduit or intermediate metal conduit. The following wiring methods shall be permitted if not subject to physical damage:

(1) Liquidtight flexible nonmetallic conduit
(2) Rigid polyvinyl chloride conduit
(3) Reinforced thermosetting resin conduit
(4) Electrical metallic tubing where installed on or within a building
(5) Electrical nonmetallic tubing where installed within a building
(6) Type MC cable where installed within a building and if not subject to corrosive environment

Exception: An existing feeder between an existing remote panelboard and service equipment shall be permitted to run in flexible metal conduit or an approved cable assembly that includes an equipment grounding conductor within its outer sheath. The equipment grounding conductor shall comply with 250.24(A)(5).

(2) Aluminum Conduit. Aluminum conduit shall not be permitted in the pool area where subject to corrosion.

(B) Grounding. An equipment grounding conductor shall be installed with the feeder conductors between the grounding terminal of the pool equipment panelboard and the grounding terminal of the applicable service equipment or source of a separately derived system. For other than (1) existing feeders covered in 680.25(A), exception, or (2) feeders to separate buildings that do not utilize an insulated equipment grounding conductor in accordance with 680.25(B)(2), this equipment grounding conductor shall be insulated.

(1) Size. This conductor shall be sized in accordance with 250.122 but not smaller than 12 AWG. On separately derived systems, this conductor shall be sized in accordance with 250.30(A)(8) but not smaller than 8 AWG.

(2) Separate Buildings. A feeder to a separate building or structure shall be permitted to supply swimming pool equipment branch circuits, or feeders supplying swimming pool equipment branch circuits, if the grounding arrangements in the separate building meet the requirements in 250.32(B). Where installed in other than existing feeders covered in 680.25(A), Exception, a separate equipment grounding conductor shall be an insulated conductor.

680.26 Equipotential Bonding.

(A) Performance. The equipotential bonding required by this section shall be installed to reduce voltage gradients in the pool area.

(B) Bonded Parts. The parts specified in 680.26(B)(1) through (B)(7) shall be bonded together using solid copper conductors, insulated covered, or bare, not smaller than 8 AWG or with rigid metal conduit of brass or other identified corrosion-resistant metal. Connections to bonded parts shall be made in accordance with 250.8. An 8 AWG or larger solid copper bonding conductor provided to reduce voltage gradients in the pool area shall not be required to be extended or attached to remote panelboards, service equipment, or electrodes.

(1) Conductive Pool Shells. Bonding to conductive pool shells shall be provided as specified in 680.26(B)(1)(a) or (B)(1)(b). Poured concrete, pneumatically applied or sprayed concrete, and concrete block with painted or plastered coatings shall all be considered conductive materials due to water permeability and porosity. Vinyl liners and fiberglass composite shells shall be considered to be nonconductive materials.

(a) *Structural Reinforcing Steel.* Unencapsulated structural reinforcing steel shall be bonded together by steel tie wires or the equivalent. Where structural reinforcing steel is encapsulated in a nonconductive compound, a copper conductor grid shall be installed in accordance with 680.26(B)(1)(b).

(b) *Copper Conductor Grid.* A copper conductor grid shall be provided and shall comply with (b)(1) through (b)(4).

(1) Be constructed of minimum 8 AWG bare solid copper conductors bonded to each other at all points of crossing. The bonding shall be in accordance with 250.8 or other approved means.
(2) Conform to the contour of the pool
(3) Be arranged in a 300-mm (12-in.) by 300-mm (12-in.) network of conductors in a uniformly spaced perpendicular grid pattern with a tolerance of 100 mm (4 in.)
(4) Be secured within or under the pool no more than 150 mm (6 in.) from the outer contour of the pool shell

(2) Perimeter Surfaces. The perimeter surface shall extend for 1 m (3 ft) horizontally beyond the inside walls of the pool and shall include unpaved surfaces, as well as poured concrete surfaces and other types of paving. Perimeter surfaces less than 1 m (3 ft) separated by a permanent wall or building 1.5 m (5 ft) in height or more shall require equipotential bonding on the pool side of the permanent wall or building. Bonding to perimeter surfaces shall be provided as specified in 680.26(B)(2)(a) or (2)(b) and shall be attached to the pool reinforcing steel or copper conductor grid at a minimum of four (4) points uniformly spaced around the perimeter of the pool. For nonconductive pool shells, bonding at four points shall not be required.

(a) *Structural Reinforcing Steel.* Structural reinforcing steel shall be bonded in accordance with 680.26(B)(1)(a).

(b) *Alternate Means.* Where structural reinforcing steel is not available or is encapsulated in a nonconductive compound, a copper conductor(s) shall be utilized where the following requirements are met:

(1) At least one minimum 8 AWG bare solid copper conductor shall be provided.

(2) The conductors shall follow the contour of the perimeter surface.

(3) Only listed splices shall be permitted.

(4) The required conductor shall be 450 mm to 600 mm (18 in. to 24 in.) from the inside walls of the pool.

(5) The required conductor shall be secured within or under the perimeter surface 100 mm to 150 mm (4 in. to 6 in.) below the subgrade.

(3) Metallic Components. All metallic parts of the pool structure, including reinforcing metal not addressed in 680.26(B)(1)(a), shall be bonded. Where reinforcing steel is encapsulated with a nonconductive compound, the reinforcing steel shall not be required to be bonded.

(4) Underwater Lighting. All metal forming shells and mounting brackets of no-niche luminaires shall be bonded.

Exception: Listed low-voltage lighting systems with nonmetallic forming shells shall not require bonding.

(5) Metal Fittings. All metal fittings within or attached to the pool structure shall be bonded. Isolated parts that are not over 100 mm (4 in.) in any dimension and do not penetrate into the pool structure more than 25 mm (1 in.) shall not require bonding.

(6) Electrical Equipment. Metal parts of electrical equipment associated with the pool water circulating system, including pump motors and metal parts of equipment associated with pool covers, including electric motors, shall be bonded.

Exception: Metal parts of listed equipment incorporating an approved system of double insulation shall not be bonded.

(a) *Double-Insulated Water Pump Motors.* Where a double-insulated water pump motor is installed under the provisions of this rule, a solid 8 AWG copper conductor of sufficient length to make a bonding connection to a replacement motor shall be extended from the bonding grid to an accessible point in the vicinity of the pool pump motor. Where there is no connection between the swimming pool bonding grid and the equipment grounding system for the premises, this bonding conductor shall be connected to the equipment grounding conductor of the motor circuit.

(b) *Pool Water Heaters.* For pool water heaters rated at more than 50 amperes and having specific instructions regarding bonding and grounding, only those parts designated to be bonded shall be bonded and only those parts designated to be grounded shall be grounded.

(7) Fixed Metal Parts. All fixed metal parts shall be bonded including, but not limited to, metal-sheathed cables and raceways, metal piping, metal awnings, metal fences, and metal door and window frames.

Exception No. 1: Those separated from the pool by a permanent barrier that prevents contact by a person shall not be required to be bonded.

Exception No. 2: Those greater than 1.5 m (5 ft) horizontally of the inside walls of the pool shall not be required to be bonded.

Exception No. 3: Those greater than 3.7 m (12 ft) measured vertically above the maximum water level of the pool, or as measured vertically above any observation stands, towers, or platforms, or any diving structures, shall not be required to be bonded.

(C) Pool Water. An intentional bond of a minimum conductive surface area of 5800 mm^2 (9 in.^{2}) shall be installed in contact with the pool water. This bond shall be permitted to consist of parts that are required to be bonded in 680.26(B).

680.27 SPECIALIZED POOL EQUIPMENT.

(A) Underwater Audio Equipment. All underwater audio equipment shall be identified for the purpose.

(B) Electrically Operated Pool Covers.

(1) Motors and Controllers. The electric motors, controllers, and wiring shall be located not less than 1.5 m (5 ft) from the inside wall of the pool unless separated from the pool by a wall, cover, or other permanent barrier. Electric motors installed below grade level shall be of the totally enclosed type. The device that controls the operation of the motor for an electrically operated pool cover shall be located such that the operator has full view of the pool.

(2) Protection. The electric motor and controller shall be connected to a circuit protected by a ground-fault circuit interrupter.

(C) Deck Area Heating. The provisions of this section shall apply to all pool deck areas, including a covered pool, where electrically operated comfort heating units are installed within 6.0 m (20 ft) of the inside wall of the pool.

(1) Unit Heaters. Unit heaters shall be rigidly mounted to the structure and shall be of the totally enclosed or guarded type. Unit heaters shall not be mounted over the pool or within the area extending 1.5 m (5 ft) horizontally from the inside walls of a pool.

(2) Permanently Wired Radiant Heaters. Radiant electric heaters shall be suitably guarded and securely fastened to their mounting device(s). Heaters shall not be installed over a pool or within the area extending 1.5 m (5 ft) horizontally from the inside walls of the pool and shall be mounted at least 3.7 m (12 ft) vertically above the pool deck unless otherwise approved.

(3) Radiant Heating Cables Not Permitted. Radiant heating cables embedded in or below the deck shall not be permitted.

III. STORABLE POOLS

680.30 General. Electrical installations at storable pools shall comply with the provisions of Part I and Part III of this article.

680.31 Pumps. A cord-connected pool filter pump shall incorporate an approved system of double insulation or its equivalent and shall be provided with means for grounding only the internal and nonaccessible non–current-carrying metal parts of the appliance.

The means for grounding shall be an equipment grounding conductor run with the power-supply conductors in the flexible cord that is properly terminated in a grounding-type attachment plug having a fixed grounding contact member.

Cord-connected pool filter pumps shall be provided with a ground-fault circuit interrupter that is an integral part of the attachment plug or located in the power supply cord within 300 mm (12 in.) of the attachment plug.

680.32 Ground-Fault Circuit Interrupters Required. All electrical equipment, including power-supply cords, used with storable pools shall be protected by ground-fault circuit interrupters.

All 125-volt, 15- and 20-ampere receptacles located within 6.0 m (20 ft) of the inside walls of a storable pool shall be protected by a ground-fault circuit interrupter. In determining these dimensions, the distance to be measured shall be the shortest path the supply cord of an appliance connected to the receptacle would follow without piercing a floor, wall, ceiling, doorway with hinged or sliding door, window opening, or other effective permanent barrier.

680.33 Luminaires. An underwater luminaire, if installed, shall be installed in or on the wall of the storable pool. It shall comply with either 680.33(A) or (B).

(A) Within the Low Voltage Contact Limit. A luminaire shall be part of a cord-and-plug-connected lighting assembly. This assembly shall be listed as an assembly for the purpose and have the following construction features:

(1) No exposed metal parts
(2) A luminaire lamp that is suitable for use at the supplied voltage
(3) An impact-resistant polymeric lens, luminaire body, and transformer enclosure
(4) A transformer or power supply meeting the requirements of 680.23(A)(2) with a primary rating not over 150 volts

(B) Over the Low Voltage Contact Limit But Not over 150 Volts. A lighting assembly without a transformer or power supply and with the luminaire lamp(s) operating at not over 150 volts shall be permitted to be cord-and-plug-connected where the assembly is listed as an assembly for the purpose. The installation shall comply with 680.23(A)(5), and the assembly shall have the following construction features:

(1) No exposed metal parts
(2) An impact-resistant polymeric lens and luminaire body
(3) A ground-fault circuit interrupter with open neutral conductor protection as an integral part of the assembly
(4) The luminaire lamp permanently connected to the ground-fault circuit interrupter with open-neutral protection
(5) Compliance with the requirements of 680.23(A)

680.34 Receptacle Locations. Receptacles shall not be located less than 1.83 m (6 ft) from the inside walls of a pool. In determining these dimensions, the distance to be measured shall be the shortest path the supply cord of an appliance connected to the receptacle would follow without piercing a floor, wall, ceiling, doorway with hinged or sliding door, window opening, or other effective permanent barrier.

IV. SPAS AND HOT TUBS

680.40 General. Electrical installations at spas and hot tubs shall comply with the provisions of Part I and Part IV of this article.

680.41 Emergency Switch for Spas and Hot Tubs. A clearly labeled emergency shutoff or control switch for the purpose of stopping the motor(s) that provide power to the recirculation system and jet system shall be installed at a point readily accessible to the users and not less than 1.5 m (5 ft) away, adjacent to, and within sight of the spa or hot tub. This requirement shall not apply to single-family dwellings.

680.42 Outdoor Installations. A spa or hot tub installed outdoors shall comply with the provisions of Parts I and II of this article, except as permitted in 680.42(A) and (B), that would otherwise apply to pools installed outdoors.

(A) Flexible Connections. Listed packaged spa or hot tub equipment assemblies or self-contained spas or hot tubs utilizing a factory-installed or assembled control panel or panelboard shall be permitted to use flexible connections as covered in 680.42(A)(1) and (A)(2).

(1) Flexible Conduit. Liquidtight flexible metal conduit or liquidtight flexible nonmetallic conduit shall be permitted in lengths of not more than 1.8 m (6 ft) external to the spa or hot tub enclosure in addition to the length needed within the enclosure to make the electrical connection.

(2) Cord-and-Plug Connections. Cord-and-plug connections with a cord not longer than 4.6 m (15 ft) shall be permitted where protected by a ground-fault circuit interrupter.

(B) Bonding. Bonding by metal-to-metal mounting on a common frame or base shall be permitted. The metal bands or hoops used to secure wooden staves shall not be required to be bonded as required in 680.26.

(C) Interior Wiring to Outdoor Installations. In the interior of a one-family dwelling or in the interior of another building or structure associated with a one-family dwelling, any of the wiring methods recognized in Chapter 3 of this *Code* that contain a copper equipment grounding conductor that is insulated or enclosed within the outer sheath of the wiring method and not smaller than 12 AWG shall be permitted to be used for the connection to motor, heating, and control loads that are part of a self-contained spa or hot tub or a packaged spa or hot tub equipment assembly. Wiring to an underwater luminaire shall comply with 680.23 or 680.33.

680.43 Indoor Installations. A spa or hot tub installed indoors shall comply with the provisions of Parts I and II of this article except as modified by this section and shall be connected by the wiring methods of Chapter 3.

Exception No. 1: Listed spa and hot tub packaged units rated 20 amperes or less shall be permitted to be cord-and-plug-connected to facilitate the removal or disconnection of the unit for maintenance and repair.

Exception No. 2: The equipotential bonding requirements for perimeter surfaces in 680.26(B)(2) shall not apply to a listed self-contained spa or hot tub installed above a finished floor.

(A) Receptacles. At least one 125-volt, 15- or 20-ampere receptacle on a general-purpose branch circuit shall be located not less than 1.83 m (6 ft) from, and not exceeding 3.0 m (10 ft) from, the inside wall of the spa or hot tub.

(1) Location. Receptacles shall be located at least 1.83 m (6 ft) measured horizontally from the inside walls of the spa or hot tub.

(2) Protection, General. Receptacles rated 125 volts and 30 amperes or less and located within 3.0 m (10 ft) of the inside walls of a spa or hot tub shall be protected by a ground-fault circuit interrupter.

(3) Protection, Spa or Hot Tub Supply Receptacle. Receptacles that provide power for a spa or hot tub shall be ground-fault circuit-interrupter protected.

(4) Measurements. In determining the dimensions in this section addressing receptacle spacings, the distance to be measured shall be the shortest path the supply cord of an appliance connected to the receptacle would follow without piercing a floor, wall, ceiling, doorway with hinged or sliding door, window opening, or other effective permanent barrier.

(B) Installation of Luminaires, Lighting Outlets, and Ceiling-Suspended (Paddle) Fans.

(1) Elevation. Luminaires, except as covered in 680.43(B)(2), lighting outlets, and ceiling-suspended (paddle) fans located over the spa or hot tub or within 1.5 m (5 ft) from the inside walls of the spa or hot tub shall comply with the clearances specified in (B)(1)(a), (B)(1)(b), and (B)(1)(c) above the maximum water level.

(a) *Without GFCI.* Where no GFCI protection is provided, the mounting height shall be not less than 3.7 m (12 ft).

(b) *With GFCI.* Where GFCI protection is provided, the mounting height shall be permitted to be not less than 2.3 m (7 ft 6 in.).

(c) *Below 2.3 m (7 ft 6 in.).* Luminaires meeting the requirements of item (1) or (2) and protected by a ground-fault circuit interrupter shall be permitted to be installed less than 2.3 m (7 ft 6 in.) over a spa or hot tub:

(1) Recessed luminaires with a glass or plastic lens, nonmetallic or electrically isolated metal trim, and suitable for use in damp locations
(2) Surface-mounted luminaires with a glass or plastic globe, a nonmetallic body, or a metallic body isolated from contact, and suitable for use in damp locations

(2) Underwater Applications. Underwater luminaires shall comply with the provisions of 680.23 or 680.33.

(C) Switches. Switches shall be located at least 1.5 m (5 ft), measured horizontally, from the inside walls of the spa or hot tub.

(D) Bonding. The following parts shall be bonded together:

(1) All metal fittings within or attached to the spa or hot tub structure

(2) Metal parts of electrical equipment associated with the spa or hot tub water circulating system, including pump motors, unless part of a listed self-contained spa or hot tub

(3) Metal raceway and metal piping that are within 1.5 m (5 ft) of the inside walls of the spa or hot tub and that are not separated from the spa or hot tub by a permanent barrier

(4) All metal surfaces that are within 1.5 m (5 ft) of the inside walls of the spa or hot tub and that are not separated from the spa or hot tub area by a permanent barrier

Exception: Small conductive surfaces not likely to become energized, such as air and water jets and drain fittings, where not connected to metallic piping, towel bars, mirror frames, and similar nonelectrical equipment, shall not be required to be bonded.

(5) Electrical devices and controls that are not associated with the spas or hot tubs and that are located less than 1.5 m (5 ft) from such units; otherwise, they shall be bonded to the spa or hot tub system

(E) Methods of Bonding. All metal parts associated with the spa or hot tub shall be bonded by any of the following methods:

(1) The interconnection of threaded metal piping and fittings

(2) Metal-to-metal mounting on a common frame or base

(3) The provisions of a solid copper bonding jumper, insulated, covered, or bare, not smaller than 8 AWG

(F) Grounding. The following equipment shall be grounded:

(1) All electrical equipment located within 1.5 m (5 ft) of the inside wall of the spa or hot tub

(2) All electrical equipment associated with the circulating system of the spa or hot tub

680.44 Protection. Except as otherwise provided in this section, the outlet(s) that supplies a self-contained spa or hot tub, a packaged spa or hot tub equipment assembly, or a field-assembled spa or hot tub shall be protected by a ground-fault circuit interrupter.

(A) Listed Units. If so marked, a listed self-contained unit or listed packaged equipment assembly that includes integral ground-fault circuit-interrupter protection for all electrical parts within the unit or assembly (pumps, air blowers, heaters, lights, controls, sanitizer generators, wiring, and so forth) shall be permitted without additional GFCI protection.

(B) Other Units. A field-assembled spa or hot tub rated 3 phase or rated over 250 volts or with a heater load of more than 50 amperes shall not require the supply to be protected by a ground-fault circuit interrupter.

V. FOUNTAINS

680.50 General. The provisions of Part I and Part V of this article shall apply to all permanently installed fountains as defined in 680.2. Fountains that have water common to a pool shall additionally comply with the requirements in Part II of this article. Part V does not cover self-contained, portable fountains. Portable fountains shall comply with Parts II and III of Article 422.

680.51 Luminaires, Submersible Pumps, and Other Submersible Equipment.

(A) Ground-Fault Circuit Interrupter. Luminaires, submersible pumps, and other submersible equipment, unless listed for operation at low voltage contact limit or less and supplied by a transformer or power supply that complies with 680.23(A)(2), shall be protected by a ground-fault circuit interrupter.

(B) Operating Voltage. No luminaires shall be installed for operation on supply circuits over 150 volts between conductors. Submersible pumps and other submersible equipment shall operate at 300 volts or less between conductors.

(C) Luminaire Lenses. Luminaires shall be installed with the top of the luminaire lens below the normal water level of the fountain unless listed for above-water locations. A luminaire facing upward shall comply with either (1) or (2):

(1) Have the lens adequately guarded to prevent contact by any person

(2) Be listed for use without a guard

(D) Overheating Protection. Electrical equipment that depends on submersion for safe operation shall be protected against overheating by a low-water cutoff or other approved means when not submerged.

(E) Wiring. Equipment shall be equipped with provisions for threaded conduit entries or be provided with a suitable flexible cord. The maximum length of each exposed cord in the fountain shall be limited to 3.0 m (10 ft). Cords extending beyond the fountain perimeter shall be enclosed in approved wiring enclosures. Metal parts of equipment in contact with water shall be of brass or other approved corrosion-resistant metal.

(F) Servicing. All equipment shall be removable from the water for relamping or normal maintenance. Luminaires shall not be permanently embedded into the fountain structure such that the water level must be reduced or the fountain drained for relamping, maintenance, or inspection.

(G) Stability. Equipment shall be inherently stable or be securely fastened in place.

680.52 Junction Boxes and Other Enclosures.

(A) General. Junction boxes and other enclosures used for other than underwater installation shall comply with 680.24.

(B) Underwater Junction Boxes and Other Underwater Enclosures. Junction boxes and other underwater enclosures shall meet the requirements of 680.52(B)(1) and (B)(2).

(1) Construction.

(a) Underwater enclosures shall be equipped with provisions for threaded conduit entries or compression glands or seals for cord entry.

(b) Underwater enclosures shall be submersible and made of copper, brass, or other approved corrosion-resistant material.

(2) Installation. Underwater enclosure installations shall comply with (a) and (b).

(a) Underwater enclosures shall be filled with an approved potting compound to prevent the entry of moisture.

(b) Underwater enclosures shall be firmly attached to the supports or directly to the fountain surface and bonded as required. Where the junction box is supported only by conduits in accordance with 314.23(E) and (F), the conduits shall be of copper, brass, stainless steel, or other approved corrosion-resistant metal. Where the box is fed by nonmetallic conduit, it shall have additional supports and fasteners of copper, brass, or other approved corrosion-resistant material.

680.53 Bonding. All metal piping systems associated with the fountain shall be bonded to the equipment grounding conductor of the branch circuit supplying the fountain.

680.54 Grounding. The following equipment shall be grounded:

(1) Other than listed low-voltage luminaires not requiring grounding, all electrical equipment located within the fountain or within 1.5 m (5 ft) of the inside wall of the fountain
(2) All electrical equipment associated with the recirculating system of the fountain
(3) Panelboards that are not part of the service equipment and that supply any electrical equipment associated with the fountain

680.55 Methods of Grounding.

(A) Applied Provisions. The provisions of 680.21(A), 680.23(B)(3), 680.23(F)(1) and (F)(2), 680.24(F), and 680.25 shall apply.

(B) Supplied by a Flexible Cord. Electrical equipment that is supplied by a flexible cord shall have all exposed non–current-carrying metal parts grounded by an insulated copper equipment grounding conductor that is an integral part of this cord. The equipment grounding conductor shall be connected to an equipment grounding terminal in the supply junction box, transformer enclosure, power supply enclosure, or other enclosure.

680.56 Cord-and-Plug-Connected Equipment.

(A) Ground-Fault Circuit Interrupter. All electrical equipment, including power-supply cords, shall be protected by ground-fault circuit interrupters.

(B) Cord Type. Flexible cord immersed in or exposed to water shall be of a type for extra-hard usage, as designated in Table 400.4, and shall be a listed type with a "W" suffix.

(C) Sealing. The end of the flexible cord jacket and the flexible cord conductor termination within equipment shall be covered with, or encapsulated in, a suitable potting compound to prevent the entry of water into the equipment through the cord or its conductors. In addition, the ground connection within equipment shall be similarly treated to protect such connections from the deteriorating effect of water that may enter into the equipment.

(D) Terminations. Connections with flexible cord shall be permanent, except that grounding-type attachment plugs and receptacles shall be permitted to facilitate removal or disconnection for maintenance, repair, or storage of fixed or stationary equipment not located in any water-containing part of a fountain.

VII. HYDROMASSAGE BATHTUBS

680.70 General. Hydromassage bathtubs as defined in 680.2 shall comply with Part VII of this article. They shall not be required to comply with other parts of this article.

680.71 Protection. Hydromassage bathtubs and their associated electrical components shall be on an individual branch circuit(s) and protected by a readily accessible ground-fault circuit interrupter. All 125-volt, single-phase receptacles not exceeding 30 amperes and located within 1.83 m (6 ft) measured horizontally of the inside walls of a hydromassage tub shall be protected by a ground-fault circuit interrupter.

680.72 Other Electrical Equipment. Luminaires, switches, receptacles, and other electrical equipment located in the same room, and not directly associated with a hydromassage bathtub, shall be installed in accordance with the requirements of Chapters 1 through 4 in this *Code* covering the installation of that equipment in bathrooms.

680.73 Accessibility. Hydromassage bathtub electrical equipment shall be accessible without damaging the building structure or building finish. Where the hydromassage bathtub is cord- and plug-connected with the supply receptacle accessible only through a service access opening, the receptacle shall be installed so that its face is within direct view and not more than 300 mm (1 ft) of the opening.

680.74 Bonding. All metal piping systems and all grounded metal parts in contact with the circulating water shall be bonded together using a solid copper bonding jumper, insulated, covered, or bare, not smaller than 8 AWG. The bonding jumper shall be connected to the terminal on the circulating pump motor that is intended for this purpose. The bonding jumper shall not be required to be connected to a double insulated circulating pump motor. The 8 AWG or larger solid copper bonding jumper shall be required for equipotential bonding in the area of the hydromassage bathtub and shall not be required to be extended or attached to any remote panelboard, service equipment, or any electrode. The 8 AWG or larger solid copper bonding jumper shall be long enough to terminate on a replacement non-double-insulated pump motor and shall be terminated to the equipment grounding conductor of the branch circuit of the motor when a double-insulated circulating pump motor is used.

CHAPTER 26

CONDUIT AND TUBING FILL TABLES

INTRODUCTION

This chapter contains tables and notes from Chapter 9—Tables and Annex C—Conduit and Tubing Fill Tables for Conductors and Fixture Wires of the Same Size. These tables are used to determine the maximum number of conductors permitted in raceways (conduit and tubing). Tables 4 and 5 of Chapter 9 are used when different size conductors are installed in the same conduit, and Annex C is used when all the conductors are the same size. Table 8—Conductor Properties (in Chapter 9) provides useful information such as size (AWG or kcmil), circular mils, overall area, and the resistance of copper and aluminum conductors.

CHAPTER 9 TABLES

Table 1 Percent of Cross Section of Conduit and Tubing for Conductors

Number of Conductors	All Conductor Types
1	53
2	31
Over 2	40

Informational Note No. 1: Table 1 is based on common conditions of proper cabling and alignment of conductors where the length of the pull and the number of bends are within reasonable limits. It should be recognized that, for certain conditions, a larger size conduit or a lesser conduit fill should be considered.

Informational Note No. 2: When pulling three conductors or cables into a raceway, if the ratio of the raceway (inside diameter) to the conductor or cable (outside diameter) is between 2.8 and 3.2, jamming can occur. While jamming can occur when pulling four or more conductors or cables into a raceway, the probability is very low.

Notes to Tables

(1) See Informative Annex C for the maximum number of conductors and fixture wires, all of the same size (total cross-sectional area including insulation) permitted in trade sizes of the applicable conduit or tubing.

(2) Table 1 applies only to complete conduit or tubing systems and is not intended to apply to sections of conduit or tubing used to protect exposed wiring from physical damage.

(3) Equipment grounding or bonding conductors, where installed, shall be included when calculating conduit or tubing fill. The actual dimensions of the equipment grounding or bonding conductor (insulated or bare) shall be used in the calculation.

(4) Where conduit or tubing nipples having a maximum length not to exceed 600 mm (24 in.) are installed between boxes, cabinets, and similar enclosures, the nipples shall be permitted to be filled to 60 percent of their total cross-sectional area, and 310.15(B)(3)(a) adjustment factors need not apply to this condition.

(5) For conductors not included in Chapter 9, such as multiconductor cables and optical fiber cables, the actual dimensions shall be used.

(6) For combinations of conductors of different sizes, use Table 5 and Table 5A for dimensions of conductors and Table 4 for the applicable conduit or tubing dimensions.

(7) When calculating the maximum number of conductors permitted in a conduit or tubing, all of the same size (total cross-sectional area including insulation), the next higher whole number shall be used to determine the maximum number of conductors permitted when the calculation results in a decimal of 0.8 or larger.

(8) Where bare conductors are permitted by other sections of this *Code,* the dimensions for bare conductors in Table 8 shall be permitted.

(9) A multiconductor cable or flexible cord of two or more conductors shall be treated as a single conductor for calculating percentage conduit fill area. For cables that have elliptical cross sections, the cross-sectional area calculation shall be based on using the major diameter of the ellipse as a circle diameter.

Table 2 Radius of Conduit and Tubing Bends

Conduit or Tubing Size		One Shot and Full Shoe Benders		Other Bends	
Metric Designator	Trade Size	mm	in.	mm	in.
16	1-1/2	101.6	4	101.6	4
21	3/4	114.3	4-1/2	127	5
27	1	146.05	5-3/4	152.4	6
35	1-1/4	184.15	7-1/4	203.2	8
41	1-1/2	209.55	8-1/2	254	10
53	2	241.3	9-1/2	304.8	12
63	2-1/2	266.7	10-1/2	381	15
78	3	330.2	13	457.2	18
91	3-1/2	381	15	533.4	21
103	4	406.4	16	609.6	24
129	5	609.6	24	762	30
155	6	762	30	914.4	36

Table 4 Dimensions and Percent Area of Conduit and Tubing
(Areas of Conduit or Tubing for the Combinations of Wires Permitted in Table 1)

Metric Designator	Trade Size	Nominal Internal Diameter		Total Area 100%		60%		1 Wire 53%		2 Wires 31%		Over 2 Wires 40%	
		mm	in.	mm²	in.²	mm²	in.²	mm²	in.²	mm²	in.²	mm²	in.²
Article 358—Electrical Metallic Tubing (EMT)													
16	1/2	15.8	0.622	196	0.304	118	0.182	104	0.161	61	0.094	78	0.122
21	3/4	20.9	0.824	343	0.533	206	0.320	182	0.283	106	0.165	137	0.213
27	1	26.6	1.049	556	0.864	333	0.519	295	0.458	172	0.268	222	0.346
35	1-1/4	35.1	1.380	968	1.496	581	0.897	513	0.793	300	0.464	387	0.598
41	1-1/2	40.9	1.610	1314	2.036	788	1.221	696	1.079	407	0.631	526	0.814
53	2	52.5	2.067	2165	3.356	1299	2.013	1147	1.778	671	1.040	866	1.342
63	2-1/2	69.4	2.731	3783	5.858	2270	3.515	2005	3.105	1173	1.816	1513	2.343
78	3	85.2	3.356	5701	8.846	3421	5.307	3022	4.688	1767	2.742	2280	3.538
91	3-1/2	97.4	3.834	7451	11.545	4471	6.927	3949	6.119	2310	3.579	2980	4.618
103	4	110.1	4.334	9521	14.753	5712	8.852	5046	7.819	2951	4.573	3808	5.901

Article 362—Electrical Nonmetallic Tubing (ENT)

16	1/2	14.2	0.560	158	0.246	95	0.148	84	0.131	49	0.076	63	0.099
21	3/4	19.3	0.760	293	0.454	176	0.272	155	0.240	91	0.141	117	0.181
27	1	25.4	1.000	507	0.785	304	0.471	269	0.416	157	0.243	203	0.314
35	1-1/4	34.0	1.340	908	1.410	545	0.846	481	0.747	281	0.437	363	0.564
41	1-1/2	39.9	1.570	1250	1.936	750	1.162	663	1.026	388	0.600	500	0.774
53	2	51.3	2.020	2067	3.205	1240	1.923	1095	1.699	641	0.993	827	1.282
63	2-1/2	—	—	—	—	—	—	—	—	—	—	—	—
78	3	—	—	—	—	—	—	—	—	—	—	—	—
91	3-1/2	—	—	—	—	—	—	—	—	—	—	—	—

Article 348—Flexible Metal Conduit (FMC)

12	3/8	9.7	0.384	74	0.116	44	0.069	39	0.061	23	0.036	30	0.046
16	1/2	16.1	0.635	204	0.317	122	0.190	108	0.168	63	0.098	81	0.127
21	3/4	20.9	0.824	343	0.533	206	0.320	182	0.283	106	0.165	137	0.213
27	1	25.9	1.020	527	0.817	316	0.490	279	0.433	163	0.253	211	0.327
35	1-1/4	32.4	1.275	824	1.277	495	0.766	437	0.677	256	0.396	330	0.511
41	1-1/2	39.1	1.538	1201	1.858	720	1.115	636	0.985	372	0.576	480	0.743
53	2	51.8	2.040	2107	3.269	1264	1.961	1117	1.732	653	1.013	843	1.307

(continued)

Table 4 Dimensions and Percent Area of Conduit and Tubing
(Areas of Conduit or Tubing for the Combinations of Wires Permitted in Table 1) *Continued*

Metric Designator	Trade Size	Nominal Internal Diameter		Total Area 100%		60%		1 Wire 53%		2 Wires 31%		Over 2 Wires 40%	
		mm	in.	mm²	in.²	mm²	in.²	mm²	in.²	mm²	in.²	mm²	in.²

Article 348—Flexible Metal Conduit (FMC) *Continued*

Metric Designator	Trade Size	mm	in.	mm²	in.²	mm²	in.²	mm²	in.²	mm²	in.²	mm²	in.²
63	2-1/2	63.5	2.500	3167	4.909	1900	2.945	1678	2.602	982	1.522	1267	1.963
78	3	76.2	3.000	4560	7.069	2736	4.241	2417	3.746	1414	2.191	1824	2.827
91	3-1/2	88.9	3.500	6207	9.621	3724	5.773	3290	5.099	1924	2.983	2483	3.848
103	4	101.6	4.000	8107	12.566	4864	7.540	4297	6.660	2513	3.896	3243	5.027

Article 342—Intermediate Metal Conduit (IMC)

Metric Designator	Trade Size	mm	in.	mm²	in.²	mm²	in.²	mm²	in.²	mm²	in.²	mm²	in.²
12	3/8	—	—	—	—	—	—	—	—	—	—	—	—
16	1/2	16.8	0.660	222	0.342	133	0.205	117	0.181	69	0.106	89	0.137
21	3/4	21.9	0.864	377	0.586	226	0.352	200	0.311	117	0.182	151	0.235
27	1	28.1	1.105	620	0.959	372	0.575	329	0.508	192	0.297	248	0.384
35	1-1/4	36.8	1.448	1064	1.647	638	0.988	564	0.873	330	0.510	425	0.659
41	1-1/2	42.7	1.683	1432	2.225	859	1.335	759	1.179	444	0.690	573	0.890

53	2	54.6	2.150	2341	3.630	1405	2.178	1241	1.924	726	1.125	937	1.452
63	2-1/2	64.9	2.557	3308	5.135	1985	3.081	1753	2.722	1026	1.592	1323	2.054
78	3	80.7	3.176	5115	7.922	3069	4.753	2711	4.199	1586	2.456	2046	3.169
91	3-1/2	93.2	3.671	6822	10.584	4093	6.351	3616	5.610	2115	3.281	2729	4.234
103	4	105.4	4.166	8725	13.631	5235	8.179	4624	7.224	2705	4.226	3490	5.452

Article 356—Liquidtight Flexible Nonmetallic Conduit (LFNC-B*)

12	3/8	12.5	0.494	123	0.192	74	0.115	65	0.102	38	0.059	49	0.077
16	1/2	16.1	0.632	204	0.314	122	0.187	108	0.166	63	0.097	81	0.125
21	3/4	21.1	0.830	350	0.541	210	0.325	185	0.287	108	0.168	140	0.216
27	1	26.8	1.054	564	0.873	338	0.524	299	0.462	175	0.270	226	0.349
35	1-1/4	35.4	1.395	984	1.528	591	0.917	522	0.810	305	0.474	394	0.611
41	1-1/2	40.3	1.588	1276	1.981	765	1.188	676	1.050	395	0.614	510	0.792
53	2	51.6	2.033	2091	3.246	1255	1.948	1108	1.720	648	1.006	836	1.298

Article 356—Liquidtight Flexible Nonmetallic Conduit (LFNC-A*)

12	3/8	12.6	0.495	125	0.192	75	0.115	66	0.102	39	0.060	50	0.077
16	1/2	16.0	0.630	201	0.312	121	0.187	107	0.165	62	0.097	80	0.125
21	3/4	21.0	0.825	346	0.535	208	0.321	184	0.283	107	0.166	139	0.214

(continued)

Table 4 Dimensions and Percent Area of Conduit and Tubing
(Areas of Conduit or Tubing for the Combinations of Wires Permitted in Table 1) *Continued*

Metric Designator	Trade Size	Nominal Internal Diameter mm	in.	Total Area 100% mm²	in.²	60% mm²	in.²	1 Wire 53% mm²	in.²	2 Wires 31% mm²	in.²	Over 2 Wires 40% mm²	in.²
Article 356—Liquidtight Flexible Nonmetallic Conduit (LFNC-A*) *Continued*													
27	1	26.5	1.043	552	0.854	331	0.513	292	0.453	171	0.265	221	0.342
35	1-1/4	35.1	1.383	968	1.502	581	0.901	513	0.796	300	0.466	387	0.601
41	1-1/2	40.7	1.603	1301	2.018	781	1.211	690	1.070	403	0.626	520	0.807
53	2	52.4	2.063	2157	3.343	1294	2.006	1143	1.772	669	1.036	863	1.337
Article 350—Liquidtight Flexible Metal Conduit (LFMC)													
12	3/8	12.5	0.494	123	0.192	74	0.115	65	0.102	38	0.059	49	0.077
16	1/2	16.1	0.632	204	0.314	122	0.188	108	0.166	63	0.097	81	0.125
21	1/4	21.1	0.830	350	0.541	210	0.325	185	0.287	108	0.168	140	0.216
27	1	26.8	1.054	564	0.873	338	0.524	299	0.462	175	0.270	226	0.349
35	1-1/4	35.4	1.395	984	1.528	591	0.917	522	0.810	305	0.474	394	0.611
41	1-1/2	40.3	1.588	1276	1.981	765	1.188	676	1.050	395	0.614	510	0.792

53	2	51.6	2.033	2091	3.246	1255	1.948	1108	1.720	648	1.006	836	1.298
63	2-1/2	63.3	2.493	3147	4.881	1888	2.929	1668	2.587	976	1.513	1259	1.953
78	3	78.4	3.085	4827	7.475	2896	4.485	2559	3.962	1497	2.317	1931	2.990
91	3-1/2	89.4	3.520	6277	9.731	3766	5.839	3327	5.158	1946	3.017	2511	3.893
103	4	102.1	4.020	8187	12.692	4912	7.615	4339	6.727	2538	3.935	3275	5.077
129	5	—	—	—	—	—	—	—	—	—	—	—	—
155	6	—	—	—	—	—	—	—	—	—	—	—	—
					Article 344—Rigid Metal Conduit								
12	3/8	16.1	0.632	204	0.314	122	0.188	108	0.166	63	0.097	81	0.125
16	1/2	21.2	0.836	353	0.549	212	0.329	187	0.291	109	0.170	141	0.220
21	3/4	27.0	1.063	573	0.887	344	0.532	303	0.470	177	0.275	229	0.355
27	1	35.4	1.394	984	1.526	591	0.916	522	0.809	305	0.473	394	0.610
35	1-1/4	41.2	1.624	1333	2.071	800	1.243	707	1.098	413	0.642	533	0.829
41	1-1/2	52.9	2.083	2198	3.408	1319	2.045	1165	1.806	681	1.056	879	1.363
53	2	63.2	2.489	3137	4.866	1882	2.919	1663	2.579	972	1.508	1255	1.946
63	2-1/2	78.5	3.090	4840	7.499	2904	4.499	2565	3.974	1500	2.325	1936	3.000
78	3	90.7	3.570	6461	10.010	3877	6.006	3424	5.305	2003	3.103	2584	4.004
91	3-1/2	102.9	4.050	8316	12.882	4990	7.729	4408	6.828	2578	3.994	3326	5.153

(continued)

Table 4 Dimensions and Percent Area of Conduit and Tubing
(Areas of Conduit or Tubing for the Combinations of Wires Permitted in Table 1) *Continued*

Metric Designator	Trade Size	Nominal Internal Diameter mm	Nominal Internal Diameter in.	Total Area 100% mm²	Total Area 100% in.²	60% mm²	60% in.²	1 Wire 53% mm²	1 Wire 53% in.²	2 Wires 31% mm²	2 Wires 31% in.²	Over 2 Wires 40% mm²	Over 2 Wires 40% in.²
						Article 344—Rigid Metal Conduit *Continued*							
129	5	128.9	5.073	13050	20.212	7830	12.127	6916	10.713	4045	6.266	5220	8.085
155	6	154.8	6.093	18812	29.158	11292	17.495	9975	15.454	5834	9.039	7528	11.663
						Article 352—Rigid PVC Conduit (PVC), Schedule 80							
12	3/8	—	—	—	—	—	—	—	—	—	—	—	—
16	1/2	13.4	0.526	141	0.217	85	0.130	75	0.115	44	0.067	56	0.087
21	3/4	18.3	0.722	263	0.409	158	0.246	139	0.217	82	0.127	105	0.164
27	1	23.8	0.936	445	0.688	267	0.413	236	0.365	138	0.213	178	0.275
35	1-1/4	31.9	1.255	799	1.237	480	0.742	424	0.656	248	0.383	320	0.495
41	1-1/2	37.5	1.476	1104	1.711	663	1.027	585	0.907	342	0.530	442	0.684
53	2	48.6	1.913	1855	2.874	1113	1.725	983	1.523	575	0.891	742	1.150
63	2-1/2	58.2	2.290	2660	4.119	1596	2.471	1410	2.183	825	1.277	1064	1.647

78	3	72.7	2.864	4151	6.442	2491	3.865	2200	3.414	1287	1.997	1660	2.577
91	3-1/2	84.5	3.326	5608	8.688	3365	5.213	2972	4.605	1738	2.693	2243	3.475
103	4	96.2	3.786	7268	11.258	4361	6.755	3852	5.967	2253	3.490	2907	4.503
129	5	121.1	4.768	11518	17.855	6911	10.713	6105	9.463	3571	5.535	4607	7.142
155	6	145.0	5.709	16513	25.598	9908	15.359	8752	13.567	5119	7.935	6605	10.239
Article 352—Rigid PVC Conduit (PVC), Schedule 40, and HDPE Conduit													
12	1/8	—	—	—	—	—	—	—	—	—	—	—	0.114
16	1/2	15.3	0.602	184	0.285	110	0.171	97	0.151	57	0.088	74	0.114
21	3/4	20.4	0.804	327	0.508	196	0.305	173	0.269	101	0.157	131	0.203
27	1	26.1	1.029	535	0.832	321	0.499	284	0.441	166	0.258	214	0.333
35	1-1/4	34.5	1.360	935	1.453	561	0.872	495	0.770	290	0.450	374	0.581
41	1-1/2	40.4	1.590	1282	1.986	769	1.191	679	1.052	397	0.616	513	0.794
53	2	52.0	2.047	2124	3.291	1274	1.975	1126	1.744	658	1.020	849	1.316
63	2-1/2	62.1	2.445	3029	4.695	1817	2.817	1605	2.488	939	1.455	1212	1.878
78	3	77.3	3.042	4693	7.268	2816	4.361	2487	3.852	1455	2.253	1877	2.907
91	3-1/2	89.4	3.521	6277	9.737	3766	5.842	3327	5.161	1946	3.018	2511	3.895
103	4	101.5	3.998	8091	12.554	4855	7.532	4288	6.654	2508	3.892	3237	5.022
129	5	127.4	5.016	12748	19.761	7649	11.856	6756	10.473	3952	6.126	5099	7.904
155	6	153.2	6.031	18433	28.567	11060	17.140	9770	15.141	5714	8.856	7373	11.427

Table 5 Dimensions of Insulated Conductors and Fixture Wires

Type: FFH-2, RFH-1, RFH-2, RHH*, RHW*, RHW-2*, RHH, RHW, RHW-2, SF-1, SF-2, SFF-1, SFF-2, TF, TFF, THHW, THW, THW-2, TW, XF, XFF

Type	Size (AWG or kcmil)	Approximate Diameter		Approximate Area	
		mm	in.	mm²	in.²
RFH-2, FFH-2	18	3.454	0.136	9.355	0.0145
	16	3.759	0.148	11.10	0.0172
RHW-2, RHH, RHW	14	4.902	0.193	18.90	0.0293
	12	5.385	0.212	22.77	0.0353
	10	5.994	0.236	28.19	0.0437
	8	8.280	0.326	53.87	0.0835
	6	9.246	0.364	67.16	0.1041
	4	10.46	0.412	86.00	0.1333
	3	11.18	0.440	98.13	0.1521
	2	11.99	0.472	112.9	0.1750
	1	14.78	0.582	171.6	0.2660
	1/0	15.80	0.622	196.1	0.3039
	2/0	16.97	0.668	226.1	0.3505
	3/0	18.29	0.720	262.7	0.4072
	4/0	19.76	0.778	306.7	0.4754

250	22.73	0.895	405.9	0.6291
300	24.13	0.950	457.3	0.7088
350	25.43	1.001	507.7	0.7870
400	26.62	1.048	556.5	0.8626
500	28.78	1.133	650.5	1.0082
600	31.57	1.243	782.9	1.2135
700	33.38	1.314	874.9	1.3561
750	34.24	1.348	920.8	1.4272
800	35.05	1.380	965.0	1.4997
900	36.68	1.444	1057	1.6377
1000	38.15	1.502	1143	1.7719
1250	43.92	1.729	1515	2.3479
1500	47.04	1.852	1738	2.6938
1750	49.94	1.966	1959	3.0357
2000	52.63	2.072	2175	3.3719
SF-2, SFF-2				
18	3.073	0.121	7.419	0.0115
16	3.378	0.133	8.968	0.0139
14	3.759	0.148	11.10	0.0172

(continued)

Table 5 Dimensions of Insulated Conductors and Fixture Wires *Continued*

Type	Size (AWG or Kcmil)	Approximate Diameter		Approximate Area	
		mm	in.	mm²	in.²
Type: FFH-2, RFH-1, RFH-2, RHH*, RHW*, RHW-2*, RHH, RHW, RHW-2, SF-1, SF-2, SFF-1, SFF-2, TF, TFF, THHW, THW, THW-2, TW, XF, XFF *Continued*					
SF-1, SFF-1	18	2.311	0.091	4.194	0.0065
RFH-1, XF, XFF	18	2.692	0.106	5.161	0.0080
TF, TFF, XF, XFF	16	2.997	0.118	7.032	0.0109
TW, XF, XFF, THHW, THW, THW-2	14	3.378	0.133	8.968	0.0139
TW, THHW, THW, THW-2	12	3.861	0.152	11.68	0.0181
	10	4.470	0.176	15.68	0.0243
	8	5.994	0.236	28.19	0.0437
RHH*, RHW*, RHW-2*	14	4.140	0.163	13.48	0.0209
RHH*, RHW*, RHW-2*, XF, XFF	12	4.623	0.182	16.77	0.0260

Type: RHH*, RHW*, RHW-2*, THHN, THHW, THW, THW-2, TFN, TFFN, THWN, THWN-2, XF, XFF					
RHH*, RHW*, RHW-2*, XF, XFF	10	5.232	0.206	21.48	0.0333
RHH*, RHW*, RHW-2*	8	6.756	0.266	35.87	0.0556
TW, THW, THHW, THW-2, RHH*, RHW*, RHW-2*	6	7.722	0.304	46.84	0.0726
	4	8.941	0.652	62.77	0.0973
	3	9.652	0.380	73.16	0.1134
	2	10.46	0.412	86.00	0.1333
	1	12.50	0.492	122.6	0.1901
	1/0	13.51	0.532	143.4	0.2223
	2/0	14.68	0.578	169.3	0.2624
	3/0	16.00	0.630	201.1	0.3117
	4/0	17.48	0.688	239.9	0.3718
	250	19.43	0.765	296.5	0.4596
	300	20.93	0.820	340.7	0.5281
	350	22.12	0.871	384.4	0.5958
	400	23.32	0.918	427.0	0.6619

(continued)

Table 5 Dimensions of Insulated Conductors and Fixture Wires *Continued*

Type	Size (AWG or kcmil)	Approximate Diameter		Approximate Area	
		mm	in.	mm²	in.²
Type: RHH*, RHW*, RHW-2*, THHN, THHW, THW, THW-2, TFN, TFFN, THWN, THWN-2, XF, XFF *Continued*					
TW, THW, THHW, THW-2, RHH*, RHW*, RHW-2*	500	25.48	1.003	509.7	0.7901
	600	28.27	1.113	627.7	0.9729
	700	30.07	1.184	710.3	1.1010
	750	30.94	1.218	751.7	1.1652
	800	31.75	1.250	791.7	1.2272
	900	33.38	1.314	874.9	1.3561
	1000	34.85	1.372	953.8	1.4784
	1250	39.09	1.539	1200	1.8602
	1500	42.21	1.662	1400	2.1695
	1750	45.11	1.776	1598	2.4773
	2000	47.80	1.882	1795	2.7818

TFN, TFFN					
	18	2.134	0.084	3.548	0.0055
	16	2.438	0.096	4.645	0.0072
THHN, THWN, THWN-2					
	14	2.819	0.111	6.258	0.0097
	12	3.302	0.130	8.581	0.0133
	10	4.166	0.164	13.61	0.0211
	8	5.486	0.216	23.61	0.0366
	6	6.452	0.254	32.71	0.0507
	4	8.230	0.324	53.16	0.0824
	3	8.941	0.352	62.77	0.0973
	2	9.754	0.384	74.71	0.1158
	1	11.33	0.446	100.8	0.1562
	1/0	12.34	0.486	119.7	0.1855
	2/0	13.51	0.532	143.4	0.2223
	3/0	14.83	0.584	172.8	0.2679
	4/0	16.31	0.642	208.8	0.3237
	250	18.06	0.711	256.1	0.3970
	300	19.46	0.766	297.3	0.4608

(continued)

Table 5 Dimensions of Insulated Conductors and Fixture Wires *Continued*

Type	Size (AWG or kcmil)	Approximate Diameter		Approximate Area	
		mm	in.	mm²	in.²
Type: FEP, FEPB, PAF, PAFF, PF PFA, PFAH, PFF, PGF, PGFF, PTF, PTFF, TFE, THHN, THWN, TWHN-2, Z, ZF, ZFF					
THHN, THWN, THWN-2	350	20.75	0.817	338.2	0.5242
	400	21.95	0.864	378.3	0.5863
	500	24.10	0.949	456.3	0.7073
	600	26.70	1.051	559.7	0.8676
	700	28.50	1.122	637.9	0.9887
	750	29.36	1.156	677.2	1.0496
	800	30.18	1.188	715.2	1.1085
	900	31.80	1.252	794.3	1.2311
	1000	33.27	1.310	869.5	1.3478
PF, PGFF, PGF, PFF, PTF, PAF, PTFF, PAFF	18	2.184	0.086	3.742	0.0058
	16	2.489	0.098	4.839	0.0075
PF, PGFF, PGF, PFF, PTF, PAF, PTFF, PAFF, TFE, FEP, PFA, FEPB, PFAH	14	2.870	0.113	6.452	0.0100

TFE, FEP, PFA, FEPB, PFAH	12	3.353	0.132	8.839	0.0137
	10	3.962	0.156	12.32	0.0191
	8	5.232	0.206	21.48	0.0333
	6	6.198	0.244	30.19	0.0468
	4	7.417	0.292	43.23	0.0670
	3	8.128	0.320	51.87	0.0804
	2	8.941	0.352	62.77	0.0973
TFE, PFAH	1	10.72	0.422	90.26	0.1399
TFE, PFA, PFAH, Z	1/0	11.73	0.462	108.1	0.1676
	2/0	12.90	0.508	130.8	0.2027
	3/0	14.22	0.560	158.9	0.2463
	4/0	15.70	0.618	193.5	0.3000
ZF, ZFF	18	1.930	0.076	2.903	0.0045
	16	2.235	0.088	3.935	0.0061
Z, ZF, ZFF	14	2.616	0.103	5.355	0.0083
Z	12	3.099	0.122	7.548	0.0117
	10	3.962	0.156	12.32	0.0191

(continued)

Table 5 Dimensions of Insulated Conductors and Fixture Wires *Continued*

Type	Size (AWG or kcmil)	Approximate Diameter		Approximate Area	
		mm	in.	mm²	in.²
Type: FEP, FEPB, PAF, PAFF, PF PFA, PFAH, PFF, PGF, PGFF, PTF, PTFF, TFE, THHN, THWN, THWN-2, Z, ZF, ZFF *Continued*						
Z	8	4.978	0.196	19.48	0.0302
	6	5.944	0.234	27.74	0.0430
	4	7.163	0.282	40.32	0.0625
	3	8.382	0.330	55.16	0.0855
	2	9.195	0.362	66.39	0.1029
	1	10.21	0.402	81.87	0.1269
Type: KF-1, KF-2, KFF-1, KFF-2, XHH, XHHW, XHHW-2, ZW						
XHHW, ZW, XHHW-2, XHH	14	3.378	0.133	8.968	0.0139
	12	3.861	0.152	11.68	0.0181
	10	4.470	0.176	15.68	0.0243
	8	5.994	0.236	28.19	0.0437
	6	6.960	0.274	38.06	0.0590
	4	8.179	0.322	52.52	0.0814
	3	8.890	0.350	62.06	0.0962
	2	9.703	0.382	73.94	0.1146

XHHW, XHHW-2, XHH

1	11.23	0.442	98.97	0.1534
1/0	12.24	0.482	117.7	0.1825
2/0	13.41	0.528	141.3	0.2190
3/0	14.73	0.580	170.5	0.2642
4/0	16.21	0.638	206.3	0.3197
250	17.91	0.705	251.9	0.3904
300	19.30	0.760	292.6	0.4536
350	20.60	0.811	333.3	0.5166
400	21.79	0.858	373.0	0.5782
500	23.95	0.943	450.6	0.6984
600	26.75	1.053	561.9	0.8709
700	28.55	1.124	640.2	0.9923
750	29.41	1.158	679.5	1.0532
800	30.23	1.190	717.5	1.1122
900	31.85	1.254	796.8	1.2351
1000	33.32	1.312	872.2	1.3519
1250	37.57	1.479	1108	1.7180
1500	40.69	1.602	1300	2.0157

(continued)

Table 5 Dimensions of Insulated Conductors and Fixture Wires *Continued*

Type	Size (AWG or kcmil)	Approximate Diameter		Approximate Area	
		mm	in.	mm²	in.²
Type: KF-1, KF-2, KFF-1, KFF-2, XHH, XHHW, XHHW-2, ZW *Continued*					
XHHW, XHHW-2, XHH	1750	43.59	1.716	1492	2.3127
	2000	46.28	1.822	1682	2.6073
KF-2, KFF-2	18	1.600	0.063	2.000	0.0031
	16	1.905	0.075	2.839	0.0044
	14	2.286	0.090	4.129	0.0064
	12	2.769	0.109	6.000	0.0093
	10	3.378	0.133	8.968	0.0139
KF-1, KFF-1	18	1.448	0.057	1.677	0.0026
	16	1.753	0.069	2.387	0.0037
	14	2.134	0.084	3.548	0.0055
	12	2.616	0.103	5.355	0.0083
	10	3.226	0.127	8.194	0.0127

*Types RHH, RHW, and RHW-2 without outer covering.

Table 8 Conductor Properties

Size (AWG or kcmil)	Conductors									Direct-Current Resistance at 75°C (167°F)					
	Area		Stranding			Overall				Copper				Aluminum	
				Diameter		Diameter		Area		Uncoated		Coated			
	mm²	Circular mils	Qty	mm	in.	mm	in.	mm²	in.²	ohm/km	ohm/kFT	ohm/km	ohm/kFT	ohm/km	ohm/kFT
18	0.823	1620	1	—	—	1.02	0.040	0.823	0.001	25.5	7.77	26.5	8.08	42.0	12.8
18	0.823	1620	7	0.39	0.015	1.16	0.046	1.06	0.002	26.1	7.95	27.7	8.45	42.8	13.1
16	1.31	2580	1	—	—	1.29	0.051	1.31	0.002	16.0	4.89	16.7	5.08	26.4	8.05
16	1.31	2580	7	0.49	0.019	1.46	0.058	1.68	0.003	16.4	4.99	17.3	5.29	26.9	8.21
14	2.08	4110	1	—	—	1.63	0.064	2.08	0.003	10.1	3.07	10.4	3.19	16.6	5.06
14	2.08	4110	7	0.62	0.024	1.85	0.073	2.68	0.004	10.3	3.14	10.7	3.26	16.9	5.17
12	3.31	6530	1	—	—	2.05	0.081	3.31	0.005	6.34	1.93	6.57	2.01	10.45	3.18
12	3.31	6530	7	0.78	0.030	2.32	0.092	4.25	0.006	6.50	1.98	6.73	2.05	10.69	3.25
10	5.261	10380	1	—	—	2.588	0.102	5.26	0.008	3.984	1.21	4.148	1.26	6.561	2.00
10	5.261	10380	7	0.98	0.308	2.95	0.116	6.76	0.011	4.070	1.24	4.226	1.29	6.679	2.04

(continued)

Table 8 Conductor Properties *Continued*

Size (AWG or kcmil)	Area mm²	Area Circular mils	Stranding Qty	Stranding Diameter mm	Stranding Diameter in.	Overall Diameter mm	Overall Diameter in.	Overall Area mm²	Overall Area in.²	Direct-Current Resistance at 75°C (167°F) Copper Uncoated ohm/km	Uncoated ohm/kFT	Coated ohm/km	Coated ohm/kFT	Aluminum ohm/km	Aluminum ohm/kFT
8	8.367	16510	1	—	—	3.264	0.128	8.37	0.013	2.506	0.764	2.579	0.786	4.125	1.26
8	8.367	16510	7	1.23	0.049	3.71	0.146	10.76	0.017	2.551	0.778	2.653	0.809	4.204	1.28
6	13.30	26240	7	1.56	0.061	4.67	0.184	17.09	0.027	1.608	0.491	1.671	0.510	2.652	0.808
4	21.15	41740	7	1.96	0.077	5.89	0.232	27.19	0.042	1.010	0.308	1.053	0.321	1.666	0.508
3	26.67	52620	7	2.20	0.087	6.60	0.260	34.28	0.053	0.802	0.245	0.833	0.254	1.320	0.403
2	33.62	66360	7	2.47	0.097	7.42	0.292	43.23	0.067	0.634	0.194	0.661	0.201	1.045	0.319
1	42.41	83690	19	1.69	0.066	8.43	0.332	55.80	0.087	0.505	0.154	0.524	0.160	0.829	0.253
1/0	53.49	105600	19	1.89	0.074	9.45	0.372	70.41	0.109	0.399	0.122	0.415	0.127	0.660	0.201
2/0	67.43	133100	19	2.13	0.084	10.62	0.418	88.74	0.137	0.3170	0.0967	0.329	0.101	0.523	0.159
3/0	85.01	167800	19	2.39	0.094	11.94	0.470	111.9	0.173	0.2512	0.0766	0.2610	0.0797	0.413	0.126
4/0	107.2	211600	19	2.68	0.106	13.41	0.528	141.1	0.219	0.1996	0.0608	0.2050	0.0626	0.328	0.100
250	—	—	37	2.09	0.082	14.61	0.575	168	0.260	0.1687	0.0515	0.1753	0.0535	0.2778	0.0847

Size														
300		37	2.29	0.090	16.00	0.630	201	0.312	0.1409	0.0429	0.1463	0.0446	0.2318	0.0707
350		37	2.47	0.097	17.30	0.681	235	0.364	0.1205	0.0367	0.1252	0.0382	0.1984	0.0605
400	—	37	2.64	0.104	18.49	0.728	268	0.416	0.1053	0.0321	0.1084	0.0331	0.1737	0.0529
500	—	37	2.95	0.116	20.65	0.813	336	0.519	0.0845	0.0258	0.0869	0.0265	0.1391	0.0424
600	—	61	2.52	0.099	22.68	0.893	404	0.626	0.0704	0.0214	0.0732	0.0223	0.1159	0.0353
700	—	61	2.72	0.107	24.49	0.964	471	0.730	0.0603	0.0184	0.0622	0.0189	0.0994	0.0303
750	—	61	2.82	0.111	25.35	0.998	505	0.782	0.0563	0.0171	0.0579	0.0176	0.0927	0.0282
800	—	61	2.91	0.114	26.16	1.030	538	0.834	0.0528	0.0161	0.0544	0.0166	0.0868	0.0265
900	—	61	3.09	0.122	27.79	1.094	606	0.940	0.0470	0.0143	0.0481	0.0147	0.0770	0.0235
1000	—	61	3.25	0.128	29.26	1.152	673	1.042	0.0423	0.0129	0.0434	0.0132	0.0695	0.0212
1250	—	91	2.98	0.117	32.74	1.289	842	1.305	0.0338	0.0103	0.0347	0.0106	0.0554	0.0169
1500	—	91	3.26	0.128	35.86	1.412	1011	1.566	0.02814	0.00858	0.02814	0.00883	0.0464	0.0141
1750	—	127	2.98	0.117	38.76	1.526	1180	1.829	0.02410	0.00735	0.02410	0.00756	0.0397	0.0121
2000	—	127	3.19	0.126	41.45	1.632	1349	2.092	0.02109	0.00643	0.02109	0.00662	0.0348	0.0106

Notes:

1. These resistance values are valid **only** for the parameters as given. Using conductors having coated strands, different stranding type, and, especially, other temperatures changes the resistance.
2. Equation for temperature change: $R_2 = R_1 [1 + \alpha(T_2 - 75)]$ where $\alpha_{cu} = 0.00323$, $\alpha_M = 0.00330$ at 75°C.
3. Conductors with compact and compressed stranding have about 9 percent and 3 percent, respectively, smaller bare conductor diameters than those shown. See Table 5A for actual compact cable dimensions.
4. The IACS conductivities used: bare copper = 100%, aluminum = 61%.
5. Class B stranding is listed as well as solid for some sizes. Its overall diameter and area is that of its circumscribing circle.

ANNEX C

Conduit and Tubing Fill Tables for Conductors and Fixture Wires of the Same Size

ANNEX C

Conduit and Tubing Fill Tables for Conductors and Fixture Wires of the Same Size

Table C1 Maximum Number of Conductors or Fixture Wires in Electrical Metallic Tubing (EMT) (Based on Table 1)

		CONDUCTORS											
					Metric Designator (Trade Size)								
Type	Conductor Size (AWG/kcmil)	16 (1/2)	21 (3/4)	27 (1)	35 (1-1/4)	41 (1-1/2)	53 (2)	63 (2-1/2)	78 (3)	91 (3-1/2)	103 (4)
RHH, RHW, RHW-2	14	4	7	11	20	27	46	80	120	157	201
	12	3	6	9	17	23	38	66	100	131	167
	10	2	5	8	13	18	30	53	81	105	135
	8	1	2	4	7	9	16	28	42	55	70
	6	1	1	3	5	8	13	22	34	44	56
	4	1	1	2	4	6	10	17	26	34	44
	3	1	1	1	4	5	9	15	23	30	38
	2	1	1	1	3	4	7	13	20	26	33
	1	0	1	1	1	3	5	9	13	17	22
	1/0	0	1	1	2	2	4	7	11	15	19
	2/0	0	1	1	2	2	4	6	10	13	17
	3/0	0	0	1	1	1	3	5	8	11	14
	4/0	0	0	1	1	1	3	5	7	9	12

250	9	7	5	3	1	1	1	0	0	0
300	8	6	5	3	1	1	1	0	0	0
350	7	6	4	3	1	1	1	0	0	0
400	7	5	4	2	1	1	1	0	0	0
500	6	4	3	2	1	1	0	0	0	0
600	5	4	3	1	1	1	0	0	0	0
700	4	3	2	1	1	0	0	0	0	0
750	4	3	2	1	1	0	0	0	0	0
800	4	3	2	1	1	0	0	0	0	0
900	3	3	1	1	1	0	0	0	0	0
1000	3	2	1	1	1	0	0	0	0	0
1250	2	1	1	1	0	0	0	0	0	0
1500	1	1	1	1	0	0	0	0	0	0
1750	1	1	1	1	0	0	0	0	0	0
2000	1	1	1	1	0	0	0	0	0	0
14	424	332	254	168	96	58	43	25	15	8
12	326	255	195	129	74	45	33	19	11	6
10	243	190	145	96	55	33	24	14	8	5
8	135	105	81	53	30	18	13	8	5	2

TW

(continued)

Table C1 Maximum Number of Conductors or Fixture Wires in Electrical Metallic Tubing (EMT) (Based on Table 1) *Continued*

CONDUCTORS

Type	Conductor Size (AWG/kcmil)	Metric Designator (Trade Size)									
		16 (1/2)	21 (3/4)	27 (1)	35 (1-1/4)	41 (1-1/2)	53 (2)	63 (2-1/2)	78 (3)	91 (3-1/2)	103 (4)
RHH*, RHW*, RHW-2*, THHW, THW, THW-2	14	6	10	16	28	39	64	112	169	221	282
RHH*, RHW*, RHW-2*, THHW, THW	12	4	8	13	23	31	51	90	136	177	227
	10	3	6	10	18	24	40	70	106	138	177
RHH*, RHW*, RHW-2*, THHW, THW, THW-2	8	1	4	6	10	14	24	42	63	83	106
RHH*, RHW*, RHW-2*, TW, THW, THHW, THW-2	6	1	3	4	8	11	18	32	48	63	81
	4	1	1	3	6	8	13	24	36	47	60
	3	1	1	3	5	7	12	20	31	40	52
	2	1	1	2	4	6	10	17	26	34	44
	1	1	1	1	3	4	7	12	18	24	31

1/0	26	20	16	10	6	3	2	1	1	0
2/0	22	17	13	9	5	3	1	1	1	0
3/0	19	15	11	7	4	2	1	1	1	0
4/0	16	12	9	6	3	1	1	1	0	0
250	13	10	7	5	3	1	1	1	0	0
300	11	8	6	4	2	1	1	1	0	0
350	10	7	6	4	1	1	1	0	0	0
400	9	7	5	3	1	1	1	0	0	0
500	7	6	4	3	1	1	1	0	0	0
600	6	4	3	2	1	1	1	0	0	0
700	5	4	3	1	1	1	0	0	0	0
750	5	4	3	1	1	1	0	0	0	0
800	5	3	3	1	1	1	0	0	0	0
900	4	3	2	1	1	0	0	0	0	0
1000	4	3	2	1	1	0	0	0	0	0
1250	3	2	1	1	1	0	0	0	0	0
1500	2	1	1	1	1	0	0	0	0	0
1750	2	1	1	1	0	0	0	0	0	0
2000	1	1	1	1	0	0	0	0	0	0

continued

Table C1 Maximum Number of Conductors or Fixture Wires in Electrical Metallic Tubing (EMT) (Based on Table 1) Continued

CONDUCTORS

Type	Conductor Size (AWG/kcmil)	Metric Designator (Trade Size)											
		16 (1/2)	21 (3/4)	27 (1)	35 (1-1/4)	41 (1-1/2)	53 (2)	63 (2-1/2)	78 (3)	91 (3-1/2)	103 (4)
THHN, THWN, THWN-2	14	12	22	35	61	84	138	241	364	476	608
	12	9	16	26	45	61	101	176	266	347	443
	10	5	10	16	28	38	63	111	167	219	279
	8	3	6	9	16	22	36	64	96	126	161
	6	2	4	7	12	16	26	46	69	91	116
	4	1	2	4	7	10	16	28	43	56	71
	3	1	1	3	6	8	13	24	36	47	60
	2	1	1	3	5	7	11	20	30	40	51
	1	1	1	1	4	5	8	15	22	29	37
	1/0	1	1	1	3	4	7	12	19	25	32
	2/0	0	1	1	2	3	6	10	16	20	26
	3/0	0	1	1	1	3	5	8	13	17	22
	4/0	0	1	1	1	2	4	7	11	14	18
	250	0	0	1	1	1	3	6	9	11	15

300	13	10	7	5	3	1	1	1	0	0
350	11	9	6	4	2	1	1	1	0	0
400	10	8	6	4	1	1	1	0	0	0
500	8	6	5	3	1	1	1	0	0	0
600	7	5	4	2	1	1	1	0	0	0
700	6	4	3	2	1	1	1	0	0	0
750	5	4	3	1	1	1	0	0	0	0
800	5	4	3	1	1	1	0	0	0	0
900	4	3	3	1	1	1	0	0	0	0
1000	4	3	2	1	1	1	0	0	0	0
14	590	462	354	234	134	81	60	34	21	12
12	430	337	258	171	98	59	43	25	15	9
10	309	241	185	122	70	42	31	18	11	6
8	177	138	106	70	40	24	18	10	6	3
6	126	98	75	50	28	17	12	7	4	2
4	88	69	53	35	20	12	9	5	3	1
3	73	57	44	29	16	10	7	4	2	1
2	60	47	36	24	13	8	6	3	1	1

FEP, FEPB, PFA, PFAH, TFE

continued

Table C1 Maximum Number of Conductors or Fixture Wires in
Electrical Metallic Tubing (EMT) (Based on Table 1) *Continued*

CONDUCTORS

Type	Conductor Size (AWG/kcmil)	Metric Designator (Trade Size)										
		16 (1/2)	21 (3/4)	27 (1)	35 (1-1/4)	41 (1-1/2)	53 (2)	63 (2-1/2)	78 (3)	91 (3-1/2)	103 (4)
PFA, PFAH, TFE	1	1	1	2	4	6	9	16	25	33	42
PFAH, TFE, PFA, PFAH, TFE, Z	1/0	1	1	1	3	5	8	14	21	27	35
	2/0	0	1	1	3	4	6	11	17	22	29
	3/0	0	1	1	2	3	5	9	14	18	24
	4/0	0	1	1	1	2	4	8	11	15	19
Z	14	14	25	41	72	98	161	282	426	556	711
	12	10	18	29	51	69	114	200	302	394	504
	10	6	11	18	31	42	70	122	185	241	309
	8	4	7	11	20	27	44	77	117	153	195
	6	3	5	8	14	19	31	54	82	107	137
	4	1	3	5	9	13	21	37	56	74	94
	3	1	2	4	7	9	15	27	41	54	69

Type	Size										
	2	57	45	34	22	13	8	6	3	1	1
	1	46	36	28	18	10	6	4	2	1	1
XHH, XHHW, XHHW-2, ZW	14	424	332	254	168	96	58	43	25	15	8
	12	326	255	195	129	74	45	33	19	11	6
	10	243	190	145	96	55	33	24	14	8	5
	8	135	105	81	53	30	18	13	8	5	2
	6	100	78	60	39	22	14	10	6	3	1
	4	72	56	43	28	16	10	7	4	2	1
	3	61	48	36	24	14	8	6	3	1	1
	2	51	40	31	20	11	7	5	3	1	1
XHH, XHHW	1	38	30	23	15	8	5	4	1	1	1
	1/0	32	25	19	13	7	4	3	1	1	1
	2/0	27	21	16	10	6	3	2	1	1	0
	3/0	22	17	13	9	5	3	1	1	1	0
	4/0	18	14	11	7	4	2	1	1	1	0
	250	15	12	9	6	3	1	1	1	0	0
	300	13	10	8	5	3	1	1	1	0	0
	350	11	9	7	4	2	1	1	0	0	0
	400	10	8	6	4	1	1	1	0	0	0

(continued)

Table C1 Maximum Number of Conductors or Fixture Wires in Electrical Metallic Tubing (EMT) (Based on Table 1) *Continued*

CONDUCTORS

Type	Conductor Size (AWG/kcmil)	Metric Designator (Trade Size)									
		16 (1/2)	21 (3/4)	27 (1)	35 (1-1/4)	41 (1-1/2)	53 (2)	63 (2-1/2)	78 (3)	91 (3-1/2)	103 (4)
XHH, XHHW, XHHW-2	500	0	0	0	1	1	1	3	5	6	8
	600	0	0	0	1	1	1	2	4	5	6
	700	0	0	0	0	1	1	2	3	4	6
	750	0	0	0	0	1	1	1	3	4	5
	800	0	0	0	0	1	1	1	3	4	5
	900	0	0	0	0	1	1	1	3	3	4
	1000	0	0	0	0	0	1	1	2	3	4
	1250	0	0	0	0	0	1	1	1	2	3
	1500	0	0	0	0	0	1	1	1	1	3
	1750	0	0	0	0	0	0	1	1	1	2
	2000	0	0	0	0	0	0	1	1	1	1

Table C2 Maximum Number of Conductors or Fixture Wires in Electrical Nonmetallic Tubing (ENT) (Based on Table 1)

Type	Conductor Size (AWG/kcmil)	CONDUCTORS Metric Designator (Trade Size)					
		16 (1/2)	21 (3/4)	27 (1)	35 (1-1/4)	41 (1-1/2)	53 (2)
RHH, RHW, RHW-2	14	3	6	10	19	26	43
	12	2	5	9	16	22	36
	10	1	4	7	13	17	29
	8	1	1	3	6	9	15
	6	1	1	3	5	7	12
	4	1	1	2	4	6	9
	3	1	1	1	3	5	8
	2	0	1	1	3	4	7
	1	0	1	1	1	3	5
	1/0	0	0	1	1	2	4
	2/0	0	0	1	1	1	3

(continued)

Table C2 Maximum Number of Conductors or Fixture Wires in Electrical Nonmetallic Tubing (ENT) (Based on Table 1) *Continued*

CONDUCTORS

Type	Conductor Size (AWG/kcmil)	Metric Designator (Trade Size)					
		16 (1/2)	21 (3/4)	27 (1)	35 (1-1/4)	41 (1-1/2)	53 (2)
RHH, RHW, RHW-2	3/0	0	0	1	1	1	3
	4/0	0	0	1	1	1	2
	250	0	0	0	1	1	1
	300	0	0	0	1	1	1
	350	0	0	0	1	1	1
	400	0	0	0	1	1	1
	500	0	0	0	0	1	1
	600	0	0	0	0	1	1
	700	0	0	0	0	0	1
	750	0	0	0	0	0	1
	800	0	0	0	0	0	1
	900	0	0	0	0	0	1

Type	Size						
	1000	0	0	0	0	0	1
	1250	0	0	0	0	0	0
	1500	0	0	0	0	0	0
	1750	0	0	0	0	0	0
	2000	0	0	0	0	0	0
TW	14	7	13	22	40	55	92
	12	5	10	17	31	42	71
	10	4	7	13	23	32	52
	8	1	4	7	13	17	29
RHH*, RHW*, RHW-2*, THHW, THW, THW-2	14	4	8	15	27	37	61
THHW, THW	12	3	7	12	21	29	49
	10	3	5	9	17	23	38
RHH*, RHW*, RHW-2*, THHW, THW, THW-2	8	1	3	5	10	14	23
RHH*, RHW*, RHW-2*, TW, THW, THHW, THW-2	6	1	2	4	7	10	17
	4	1	1	3	5	8	13
	3	1	1	2	5	7	11

(continued)

Table C2 Maximum Number of Conductors or Fixture Wires in Electrical Nonmetallic Tubing (ENT) (Based on Table 1) *Continued*

CONDUCTORS

Type	Conductor Size (AWG/kcmil)	Metric Designator (Trade Size)					
		16 (1/2)	21 (3/4)	27 (1)	35 (1-1/4)	41 (1-1/2)	53 (2)
RHH*, RHW*, RHW-2*, TW, THW, THHW, THW-2	2	1	1	2	4	6	9
	1	0	1	1	3	4	6
	1/0	0	1	1	2	3	5
	2/0	0	0	1	1	3	5
	3/0	0	0	1	1	2	4
	4/0	0	0	1	1	1	3
	250	0	0	1	0	1	2
	300	0	0	0	1	1	2
	350	0	0	0	1	1	1
	400	0	0	0	1	1	1
	500	0	0	0	1	1	1
	600	0	0	0	0	1	1
	700	0	0	0	0	1	1

THHN, THWN, THWN-2

Size						
750	0	0	0	0	1	1
800	0	0	0	0	1	1
900	0	0	0	0	0	1
1000	0	0	0	0	0	1
1250	0	0	0	0	0	1
1500	0	0	0	0	0	0
1750	0	0	0	0	0	0
2000	0	0	0	0	0	0
14	10	18	32	58	80	132
12	7	13	23	42	58	96
10	4	8	15	26	36	60
8	2	5	8	15	21	35
6	1	3	6	11	15	25
4	1	1	4	7	9	15
3	1	1	3	5	8	13
2	1	1	2	5	6	11
1	1	1	1	3	5	8
1/0	0	1	1	3	4	7
2/0	0	1	1	2	3	5

(continued)

Table C2 Maximum Number of Conductors or Fixture Wires in Electrical Nonmetallic Tubing (ENT) (Based on Table 1) *Continued*

CONDUCTORS

Type	Conductor Size (AWG/kcmil)	Metric Designator (Trade Size)						
		16 (1/2)	21 (3/4)	27 (1)	35 (1-1/4)	41 (1-1/2)	53 (2)
THHN, THWN, THWN-2	3/0	0	1	1	1	3	4
	4/0	0	0	1	1	2	4
	250	0	0	1	1	1	3
	300	0	0	1	1	1	2
	350	0	0	0	1	1	2
	400	0	0	0	1	1	1
	500	0	0	0	1	1	1
	600	0	0	0	1	1	1
	700	0	0	0	0	1	1
	750	0	0	0	0	1	1
	800	0	0	0	0	1	1
	900	0	0	0	0	1	1
	1000	0	0	0	0	0	1

FEP, FEPB, PFA, PFAH, TFE	14	10	18	31	56	77	128	
	12	7	13	23	41	56	93	
	10	5	9	16	29	40	67	
	8	3	5	9	17	23	38	
	6	1	4	6	12	16	27	
	4	1	2	4	8	11	19	
	3	1	1	4	7	9	16	
	2	1	1	3	5	8	13	
PFA, PFAH, TFE	1	1	1	1	4	5	9	
PFA, PFAH, TFE, Z	1/0	0	1	1	3	4	7	
	2/0	0	1	1	2	4	6	
	3/0	0	1	1	1	3	5	
	4/0	0	1	1	1	2	4	
Z	14	12	22	38	68	93	154	
	12	8	15	27	48	66	109	

(continued)

Table C2 Maximum Number of Conductors or Fixture Wires in Electrical Nonmetallic Tubing (ENT) (Based on Table 1) *Continued*

CONDUCTORS

Type	Conductor Size (AWG/kcmil)	Metric Designator (Trade Size)						
		16 (1/2)	21 (3/4)	27 (1)	35 (1-1/4)	41 (1-1/2)	53 (2)
Z	10	5	9	16	29	40	67
	8	3	6	10	18	25	42
	6	1	4	7	13	18	30
	4	1	3	5	9	12	20
	3	1	1	3	6	9	15
	2	1	1	3	5	7	12
	1	1	1	2	4	6	10
XHH, XHHW, XHHW-2, ZW	14	7	13	22	40	55	92
	12	5	10	17	31	42	71
	10	4	7	13	23	32	52
	8	1	4	7	13	17	29
	6	1	3	5	9	13	21

XHH, XHHW, XHHW-2, ZW						
4	1	1	4	7	9	15
3	1	1	3	6	8	13
2	1	1	2	5	6	11
XHH, XHHW, XHHW-2						
1	1	1	1	3	5	8
1/0	0	1	1	3	4	7
2/0	0	1	1	2	3	6
3/0	0	1	1	1	3	5
4/0	0	0	1	1	2	4
250	0	0	1	1	1	3
300	0	0	1	1	1	3
350	0	0	1	1	1	2
400	0	0	0	1	1	1
500	0	0	0	0	1	1
600	0	0	0	1	1	1
700	0	0	0	0	1	1
750	0	0	0	0	1	1
800	0	0	0	0	1	1
900	0	0	0	0	1	1

(continued)

Table C2 Maximum Number of Conductors or Fixture Wires in Electrical Nonmetallic Tubing (ENT) (Based on Table 1) *Continued*

CONDUCTORS

Type	Conductor Size (AWG/kcmil)	Metric Designator (Trade Size)						
		16 (1/2)	21 (3/4)	27 (1)	35 (1-1/4)	41 (1-1/2)	53 (2)
XHH, XHHW, XHHW-2	1000	0	0	0	0	0	1
	1250	0	0	0	0	0	1
	1500	0	0	0	0	0	1
	1750	0	0	0	0	0	0
	2000	0	0	0	0	0	0

Table C3 Maximum Number of Conductors or Fixture Wires in Flexible Metal Conduit (FMC) (Based on Table 1)

Type	Conductor Size (AWG/ kcmil)	CONDUCTORS Metric Designator (Trade Size)										
		16 (1/2)	21 (3/4)	27 (1)	35 (1-1/4)	41 (1-1/2)	53 (2)	63 (2-1/2)	78 (3)	91 (3-1/2)	103 (4)
RHH, RHW, RHW-2	14	4	7	11	17	25	44	67	96	131	171
	12	3	6	9	14	21	37	55	80	109	142
	10	3	5	7	11	17	30	45	64	88	115
	8	1	2	4	6	9	15	23	34	46	60
	6	1	1	3	5	7	12	19	27	37	48
	4	1	1	2	4	5	10	14	21	29	37
	3	1	1	1	3	5	8	13	18	25	33
	2	1	1	1	3	4	7	11	16	22	28
	1	0	1	1	1	2	5	7	10	14	19
	1/0	0	1	1	1	2	4	6	9	12	16
	2/0	0	1	1	1	1	3	5	8	11	14

(continued)

Table C3 Maximum Number of Conductors or Fixture Wires in Flexible Metal Conduit (FMC) (Based on Table 1) *Continued*

		CONDUCTORS										
		Metric Designator (Trade Size)										
Type	Conductor Size (AWG/kcmil)	16 (1/2)	21 (3/4)	27 (1)	35 (1-1/4)	41 (1-1/2)	53 (2)	63 (2-1/2)	78 (3)	91 (3-1/2)	103 (4)
RHH, RHW, RHW-2	3/0	0	0	1	1	1	3	3	7	9	12
	4/0	0	0	1	1	1	2	4	6	8	10
	250	0	0	0	1	1	1	3	4	6	8
	300	0	0	0	1	1	1	2	4	5	7
	350	0	0	0	0	1	1	2	3	5	6
	400	0	0	0	0	1	1	1	3	4	6
	500	0	0	0	0	1	1	1	3	4	5
	600	0	0	0	0	0	1	1	2	3	4
	700	0	0	0	0	0	1	1	1	3	3
	750	0	0	0	0	0	1	1	1	2	3
	800	0	0	0	0	0	1	1	1	2	3
	900	0	0	0	0	0	1	1	1	2	3

Type	Size										
	1000	0	0	0	0	0	1	1	1	1	3
	1250	0	0	0	0	0	0	1	1	1	1
	1500	0	0	0	0	0	0	1	1	1	1
	1750	0	0	0	0	0	0	1	1	1	1
	2000	0	0	0	0	0	0	0			
TW	14	9	15	23	36	53	94	141	203	277	361
	12	7	11	18	28	41	72	108	156	212	277
	10	5	8	13	21	30	54	81	116	158	207
	8	3	5	7	11	17	30	45	64	88	115
RHH*, RHW*, RHW-2*, THHW, THW, THW-2	14	6	10	15	24	35	62	94	135	184	240
RHH*, RHW*, RHW-2*, THHW, THW	12	5	8	12	19	28	50	75	108	148	193
	10	4	6	10	15	22	39	59	85	115	151
RHH*, RHW*, RHW-2*, THHW, THW, THW-2	8	1	4	6	9	13	23	35	51	69	90

(continued)

Table C3 Maximum Number of Conductors or Fixture Wires in Flexible Metal Conduit (FMC) (Based on Table 1) *Continued*

CONDUCTORS

Type	Conductor Size (AWG/kcmil)	Metric Designator (Trade Size)									
		16 (1/2)	21 (3/4)	27 (1)	35 (1-1/4)	41 (1-1/2)	53 (2)	63 (2-1/2)	78 (3)	91 (3-1/2)	103 (4)
RHH*, RHW*, RHW-2*, TW, THW, THHW, THW-2	6	1	3	4	7	10	18	27	39	53	69
	4	1	1	3	5	7	13	20	29	39	51
	3	1	1	3	4	6	11	17	25	34	44
	2	1	1	2	4	5	10	14	21	29	37
	1	1	1	1	2	4	7	10	15	20	26
	1/0	0	1	1	1	3	6	9	12	17	22
	2/0	0	1	1	1	3	5	7	10	14	19
	3/0	0	1	1	1	2	4	6	9	12	16
	4/0	0	0	1	1	1	3	5	7	10	13
	250	0	0	1	1	1	3	4	6	8	11
	300	0	0	1	1	1	2	3	5	7	9
	350	0	0	0	1	1	1	3	4	6	8

400	0	0	0	1	1	1	3	4	6	7
500	0	0	0	1	1	1	2	3	5	6
600	0	0	0	0	1	1	1	3	4	5
700	0	0	0	0	1	1	1	2	3	4
750	0	0	0	0	1	1	1	2	3	4
800	0	0	0	0	1	1	1	1	3	4
900	0	0	0	0	0	1	1	1	3	3
1000	0	0	0	0	0	1	1	1	2	3
1250	0	0	0	0	0	1	1	1	1	2
1500	0	0	0	0	0	0	1	1	1	1
1750	0	0	0	0	0	0	1	1	1	1
2000	0	0	0	0	0	0	1	1	1	1
THHN, THWN, THWN-2										
14	13	22	33	52	76	134	202	291	396	518
12	9	16	24	38	56	98	147	212	289	378
10	6	10	15	24	35	62	93	134	182	238
8	3	6	9	14	20	35	53	77	105	137
6	2	4	6	10	14	25	38	55	76	99

(continued)

Table C3 Maximum Number of Conductors or Fixture Wires
in Flexible Metal Conduit (FMC) (Based on Table 1) *Continued*

CONDUCTORS

Type	Conductor Size (AWG/kcmil)	Metric Designator (Trade Size)										
		16 (1/2)	21 (3/4)	27 (1)	35 (1-1/4)	41 (1-1/2)	53 (2)	63 (2-1/2)	78 (3)	91 (3-1/2)	103 (4)
THHN, THWN, THWN-2	4	1	2	4	6	9	16	24	34	46	61
	3	1	1	3	5	7	13	20	29	39	51
	2	1	1	3	4	6	11	17	24	33	43
	1	1	1	1	3	4	8	12	18	24	32
	1/0	1	1	1	2	4	7	10	15	20	27
	2/0	0	1	1	1	3	6	9	12	17	22
	3/0	0	1	1	1	2	5	7	10	14	18
	4/0	0	1	1	1	1	4	6	8	12	15
	250	0	0	1	1	1	3	5	7	9	12
	300	0	0	1	1	1	3	4	6	8	11
	350	0	0	1	1	1	2	3	5	7	9
	400	0	0	0	1	1	1	3	5	6	8
	500	0	0	0	1	1	1	2	4	5	7

Type	Size										
	600	0	0	0	0	1	1	1	3	4	5
	700	0	0	0	0	1	1	1	3	4	5
	750	0	0	0	0	1	1	1	2	3	4
	800	0	0	0	0	0	1	1	2	3	4
	900	0	0	0	0	0	1	1	1	3	4
	1000	0	0	0	0	0	1	1	1	3	3
FEP, FEPB, PFA, PFAH, TFE	14	12	21	32	51	74	130	196	282	385	502
	12	9	15	24	37	54	95	143	206	281	367
	10	6	11	17	26	39	68	103	148	201	263
	8	4	6	10	15	22	39	59	85	115	151
	6	2	4	7	11	16	28	42	60	82	107
	4	1	3	5	7	11	19	29	42	57	75
	3	1	2	4	6	9	16	24	35	48	62
	2	1	1	3	5	7	13	20	29	39	51
PFA, PFAH, TFE	1	1	1	2	3	5	9	14	20	27	36
PFA, PFAH, TFE, Z	1/0	1	1	1	3	4	8	11	17	23	30
	2/0	1	1	1	2	3	6	9	14	19	24

(continued)

Table C3 Maximum Number of Conductors or Fixture Wires in Flexible Metal Conduit (FMC) (Based on Table 1) *Continued*

CONDUCTORS

Type	Conductor Size (AWG/kcmil)	Metric Designator (Trade Size)									
		16 (1/2)	21 (3/4)	27 (1)	35 (1-1/4)	41 (1-1/2)	53 (2)	63 (2-1/2)	78 (3)	91 (3-1/2)	103 (4)
PFA, PFAH, TFE, Z	3/0	0	1	1	1	3	5	8	11	15	20
	4/0	0	1	1	1	2	4	6	9	13	16
Z	14	15	25	39	61	89	157	236	340	463	605
	12	11	18	28	43	63	111	168	241	329	429
	10	6	11	17	26	39	68	103	148	201	263
	8	4	7	11	17	24	43	65	93	127	166
	6	3	5	7	12	17	30	45	65	89	117
	4	1	3	5	8	12	21	31	45	61	80
	3	1	2	4	6	8	15	23	33	45	58
	2	1	1	3	5	7	12	19	27	37	49
	1	1	1	2	4	6	10	15	22	30	39

Type	Size										
XHH, XHHW, XHHW-2, ZW	14	9	15	23	36	53	94	141	203	277	361
	12	7	11	18	28	41	72	108	156	212	277
	10	5	8	13	21	30	54	81	116	158	207
	8	3	5	7	11	17	30	45	64	88	115
	6	1	3	5	8	12	22	33	48	65	85
	4	1	2	4	6	9	16	24	34	47	61
	3	1	1	3	5	7	13	20	29	40	52
	2	1	1	3	4	6	11	17	24	33	44
XHH, XHHW, XHHW-2	1	1	1	1	3	5	8	13	18	25	32
	1/0	1	1	1	2	4	7	10	15	21	27
	2/0	0	1	1	2	3	6	9	13	17	23
	3/0	0	1	1	1	3	5	7	10	14	19
	4/0	0	0	1	1	2	4	6	9	12	15
XHH, XHHW-2	250	0	0	1	1	1	3	5	7	10	13
	300	0	0	1	1	1	3	4	6	8	11
	350	0	0	1	1	2	2	4	5	7	9
	400	0	0	0	1	1	1	3	5	6	8
	500	0	0	0	1	1	1	3	4	5	7

(continued)

Table C3 Maximum Number of Conductors or Fixture Wires in Flexible Metal Conduit (FMC) (Based on Table 1) Continued

Type	Conductor Size (AWG/kcmil)	CONDUCTORS Metric Designator (Trade Size)									
		16 (1/2)	21 (3/4)	27 (1)	35 (1-1/4)	41 (1-1/2)	53 (2)	63 (2-1/2)	78 (3)	91 (3-1/2)	103 (4)
XHH, XHHW, XHHW-2	600	0	0	0	0	1	1	1	3	4	5
	700	0	0	0	0	1	1	1	3	4	5
	750	0	0	0	0	1	1	1	2	3	4
	800	0	0	0	0	1	1	1	2	3	4
	900	0	0	0	0	0	1	1	1	3	4
	1000	0	0	0	0	0	1	1	1	3	3
	1250	0	0	0	0	0	1	1	1	1	3
	1500	0	0	0	0	0	0	1	1	1	2
	1750	0	0	0	0	0	0	1	1	1	1
	2000	0	0	0	0	0	0	1	1	1	1

*Types RHH, RHW, and RHW-2 without outer covering.

Table C4 Maximum Number of Conductors or Fixture Wires in Intermediate Metal Conduit (IMC) (Based on Table 1)

	Conductor Size (AWG/ kcmil)	Metric Designator (Trade Size)									
Type		16 (1/2)	21 (3/4)	27 (1)	35 (1-1/4)	41 (1-1/2)	53 (2)	63 (2-1/2)	78 (3)	91 (3-1/2)	103 (4)
RHH, RHW, RHW-2	14	4	8	13	22	30	49	70	108	144	186
	12	4	6	11	18	25	41	58	89	120	154
	10	3	5	8	15	20	33	47	72	97	124
	8	1	3	4	8	10	17	24	38	50	65
	6	1	1	3	6	8	14	19	30	40	52
	4	1	1	3	5	6	11	15	23	31	41
	3	1	1	2	4	6	9	13	21	28	36
	2	0	1	1	2	5	8	11	18	24	31
	1	0	1	1	2	3	5	7	12	16	20
	1/0	0	1	1	1	3	4	6	10	14	18
	2/0	0	1	1	1	2	4	6	9	12	15

CONDUCTORS

(continued)

Table C4 Maximum Number of Conductors or Fixture Wires in Intermediate Metal Conduit (IMC) (Based on Table 1) *Continued*

CONDUCTORS

Type	Conductor Size (AWG/ kcmil)	16 (1/2)	21 (3/4)	27 (1)	35 (1-1/4)	41 (1-1/2)	53 (2)	63 (2-1/2)	78 (3)	91 (3-1/2)	103 (4)
						Metric Designator (Trade Size)					
RHH, RHW, RHW-2	3/0	0	0	1	1	1	3	5	7	10	13
	4/0	0	0	1	1	1	3	4	6	9	11
	250	0	0	1	1	1	1	3	5	6	8
	300	0	0	0	1	1	1	3	4	6	7
	350	0	0	0	1	1	1	2	4	5	7
	400	0	0	0	1	1	1	2	3	5	6
	500	0	0	0	1	1	1	1	3	4	5
	600	0	0	0	0	0	1	1	2	3	4
	700	0	0	0	0	1	1	1	2	3	4
	750	0	0	0	0	1	1	1	1	3	4
	800	0	0	0	0	0	1	1	1	2	3
	900	0	0	0	0	0	1	1	1	2	3
	1000	0	0	0	0	0	1	1	1	2	3
	1250	0	0	0	0	1	1	1	1	1	2

Type	Size										
	1500	0	0	0	0	0	0	1	1	1	1
	1750	0	0	0	0	0	0	1	1	1	1
	2000	0	0	0	0	0	0	1	1	1	1
TW	14	10	17	27	47	64	104	147	228	304	392
	12	7	13	21	36	49	80	113	175	234	301
	10	5	9	15	27	36	59	84	130	174	224
	8	3	5	8	15	20	33	47	72	97	124
RHH*, RHW*, RHW-2*, THHW, THW, THW-2	14	6	11	18	31	42	69	98	151	202	261
RHH*, RHW*, RHW-2* THHW, THW	12	5	9	14	25	34	56	79	122	163	209
	10	4	7	11	19	26	43	61	95	127	163
RHH*, RHW*- RHW-2*, THHW, THW, THW-2	8	2	4	7	12	16	26	37	57	76	98
RHH*, RHW*, RHW-2* TW, THW, THHW, THW-2	6	1	3	5	9	12	20	28	43	58	75
	4	1	2	4	6	9	15	21	32	43	56

(continued)

Table C4 Maximum Number of Conductors or Fixture Wires In Intermediate Metal Conduit (IMC) (Based on Table 1) *Continued*

CONDUCTORS

Type	Conductor Size (AWG/kcmil)	Metric Designator (Trade Size)									
		16 (1/2)	21 (3/4)	27 (1)	35 (1-1/4)	41 (1-1/2)	53 (2)	63 (2-1/2)	78 (3)	91 (3-1/2)	103 (4)
TW, THW, THHW, THW-2, TH, THW, THHW, THW-2	3	1	1	3	6	8	13	18	28	37	48
	2	1	1	3	5	6	11	15	23	31	41
	1	1	1	1	3	4	7	11	16	22	28
	1/0	1	1	1	3	4	6	9	14	19	24
	2/0	0	1	1	2	3	5	8	12	16	20
	3/0	0	1	1	1	3	4	6	10	13	17
	4/0	0	1	1	1	2	4	5	8	11	14
	250	0	0	1	1	1	3	4	7	9	12
	300	0	0	1	1	1	2	4	6	8	10
	350	0	0	1	1	1	2	3	5	7	9
	400	0	0	0	1	1	1	3	4	6	8
	500	0	0	0	1	1	1	2	4	5	7

Size										
600	0	0	0	1	1	1	1	3	4	5
700	0	0	0	0	1	1	1	3	4	5
750	0	0	0	0	1	1	1	2	3	4
800	0	0	0	0	1	1	1	2	3	4
900	0	0	0	0	1	1	1	2	3	4
1000	14	24	39	68	91	149	211	326	436	562
1250	10	17	29	49	67	109	154	238	318	410
1500	6	11	18	31	42	68	97	150	200	258
1750	3	6	10	18	24	39	56	86	115	149
2000	2	4	7	13	17	28	40	62	83	107

THHN, THWN, THWN-2

Wire sizes (AWG): 14, 12, 10, 8, 6 / 4, 3

Size										
4	1	3	4	8	10	17	25	38	51	66
3	1	2	4	6	9	15	21	32	43	56

(continued)

Table C4 Maximum Number of Conductors or Fixture Wires in Intermediate Metal Conduit (IMC) (Based on Table 1) *Continued*

CONDUCTORS

Type	Conductor Size (AWG/kcmil)	Metric Designator (Trade Size)									
		16 (1/2)	21 (3/4)	27 (1)	35 (1-1/4)	41 (1-1/2)	53 (2)	63 (2-1/2)	78 (3)	91 (3-1/2)	103 (4)
THHN, THWN, THWN-2	2	1	1	3	5	7	12	17	27	36	47
	1	1	1	2	4	5	9	13	20	27	35
	1/0	1	1	1	3	4	8	11	17	23	29
	2/0	1	1	1	2	4	6	9	14	19	24
	3/0	0	1	1	2	3	5	7	12	16	20
	4/0	0	1	1	1	2	4	6	9	13	17
	250	0	0	1	1	1	3	5	8	10	13
	300	0	0	1	1	1	3	4	7	9	12
	350	0	0	1	1	1	2	4	6	8	10
	400	0	0	0	1	1	2	3	5	7	9
	500	0	0	0	1	1	1	3	4	6	7

Type	Size										
	600	0	0	0	1	1	1	2	3	5	6
	700	0	0	0	1	1	1	1	3	4	5
	750	0	0	0	1	1	1	1	3	4	5
	800	0	0	0	0	1	1	1	3	4	5
	900	0	0	0	0	1	1	1	2	3	4
	1000	0	0	0	0	1	1	1	2	3	4
FEP, FEPB, PFA, PFAH, TFE	14	13	23	38	66	89	145	205	317	423	545
	12	10	17	28	48	65	106	150	231	309	398
	10	7	12	20	34	46	76	107	166	221	285
	8	4	7	11	19	26	43	61	95	127	163
	6	3	5	8	14	19	31	44	67	90	116
	4	1	3	5	10	13	21	30	47	63	81
	3	1	3	4	8	11	18	25	39	52	68
	2	1	2	4	6	9	15	21	32	43	56
PFA, PFAH, TFE	1	1	1	2	4	6	10	14	22	30	39
PFA, PFAH, TFE, Z	1/0	1	1	1	4	5	8	12	19	25	32
	2/0	1	1	1	3	4	7	10	15	21	27

(continued)

Table C4 Maximum Number of Conductors or Fixture Wires
in Intermediate Metal Conduit (IMC) (Based on Table 1) *Continued*

Type	Conductor Size (AWG/kcmil)	CONDUCTORS Metric Designator (Trade Size)											
		16 (1/2)	21 (3/4)	27 (1)	35 (1-1/4)	41 (1-1/2)	53 (2)	63 (2-1/2)	78 (3)	91 (3-1/2)	103 (4)
PFA, PFAH, TFE, Z	3/0	0	1	1	2	3	6	8	13	17	22
	4/0	0	1	1	1	3	5	7	10	14	18
Z	14	16	28	46	79	107	175	247	381	510	657
	12	11	20	32	56	76	124	175	271	362	466
	10	7	12	20	34	46	76	107	166	221	285
	8	4	7	12	21	29	48	68	105	140	180
	6	3	5	9	15	20	33	47	73	98	127
	4	1	3	6	10	14	23	33	50	67	87
	3	1	2	4	7	10	17	24	37	49	63
	2	1	1	3	6	8	14	20	30	41	53
	1	1	1	3	5	7	11	16	25	33	43

Type	Size										
XHH, XHHW, XHHW-2, ZW	14	10	17	27	47	64	104	147	228	304	392
	12	7	13	21	36	49	80	113	175	234	301
	10	5	9	15	27	36	59	84	130	174	224
	8	3	5	8	15	20	33	47	72	97	124
	6	1	4	6	11	15	24	35	53	71	92
	4	1	3	4	8	11	18	25	39	52	67
	3	1	2	4	7	9	15	21	33	44	56
	2	1	1	3	5	7	12	18	27	37	47
XHH, XHHW-2	1	1	1	2	4	5	9	13	20	27	35
	1/0	1	1	1	3	5	8	11	17	23	30
	2/0	1	1	1	3	4	6	9	14	19	25
	3/0	0	1	1	2	3	5	7	12	16	20
	4/0	0	1	1	1	2	4	6	10	13	17
	250	0	0	1	1	1	3	5	8	11	14
	300	0	0	1	1	1	3	4	7	9	12
	350	0	0	1	1	1	3	4	6	8	10
	400	0	0	1	1	1	2	3	5	7	9
	500	0	0	0	1	1	1	3	4	6	8

(continued)

Table C4 Maximum Number of Conductors or Fixture Wires
in Intermediate Metal Conduit (IMC) (Based on Table 1) *Continued*

	CONDUCTORS									
	Conductor Size (AWG/kcmil)	Metric Designator (Trade Size)								
Type		16 (1/2)	21 (3/4)	27 (1)	35 (1-1/4)	41 (1-1/2)	53 (2)	63 (2-1/2)	78 (3)	91 (3-1/2)	103 (4)
XHH, XHHW, XHHW-2	600	0	0	0	0	1	1	2	3	5	6
	700	0	0	0	1	1	1	1	3	4	5
	750	0	0	0	1	1	1	1	3	4	5
	800	0	0	0	0	1	1	1	3	4	5
	900	0	0	0	0	1	1	1	2	3	4
	1000	0	0	0	0	1	1	1	2	3	4
	1250	0	0	0	0	0	1	1	1	2	3
	1500	0	0	0	0	0	1	1	1	1	2
	1750	0	0	0	0	0	1	1	1	1	2
	2000	0	0	0	0	0	0	1	1	1	1

Table C5 Maximum Number of Conductors or Fixture Wires in Liquidtight Flexible Nonmetallic Conduit (Type LFNC-B*) (Based on Table 1)

	CONDUCTORS							
		Metric Designator (Trade Size)						
Type	Conductor Size (AWG/kcmil)	12 (3/8)	16 (1/2)	21 (3/4)	27 (1)	35 (1-1/4)	41 (1-1/2)	53 (2)
RHH, RHW, RHW-2	14	2	4	7	12	21	27	44
	12	1	3	6	10	17	22	36
	10	1	3	5	8	14	18	29
	8	1	1	2	4	7	9	15
	6	1	1	1	3	6	7	12
	4	0	1	1	2	4	6	9
	3	0	1	1	1	4	5	8
	2	0	1	1	1	3	4	7
	1	0	0	1	1	1	3	5
	1/0	0	0	1	1	1	2	4
	2/0	0	0	1	1	1	1	3

(continued)

Table C5 Maximum Number of Conductors or Fixture Wires in Liquidtight Flexible
Nonmetallic Conduit (Type LFNC-B*) (Based on Table 1) *Continued*

CONDUCTORS

Type	Conductor Size (AWG/kcmil)	Metric Designator (Trade Size)							
		12 (3/8)	16 (1/2)	21 (3/4)	27 (1)	35 (1-1/4)	41 (1-1/2)	53 (2)
RHH, RHW, RHW-2	3/0	0	0	0	1	1	1	3
	4/0	0	0	0	1	1	1	2
	250	0	0	0	0	1	1	1
	300	0	0	0	0	1	1	1
	350	0	0	0	0	1	1	1
	400	0	0	0	0	1	1	1
	500	0	0	0	0	1	1	1
	600	0	0	0	0	0	1	1
	700	0	0	0	0	0	0	1
	750	0	0	0	0	0	0	1
	800	0	0	0	0	0	0	1
	900	0	0	0	0	0	0	1

Type	Size							
TW	1000	0	0	0	0	0	0	1
	1250	0	0	0	0	0	0	0
	1500	0	0	0	0	0	0	0
	1750	0	0	0	0	0	0	0
	2000	0	0	0	0	0	0	0
	14	5	9	15	25	44	57	93
	12	4	7	12	19	33	43	71
	10	3	5	9	14	25	32	53
	8	1	3	5	8	14	18	29
RHH†, RHW†, RHW-2†, THW, THW-2	14	3	6	10	16	29	38	62
RHH†, RHW†, RHW-2†, THW	12	3	5	8	13	23	30	50
	10	1	3	6	10	18	23	39
RHH†, RHW†, RHW-2†, THW, THW-2	8	1	1	4	6	11	14	23

(continued)

Table C5 Maximum Number of Conductors or Fixture Wires in Liquidtight Flexible Nonmetallic Conduit (Type LFNC-B*) (Based on Table 1) *Continued*

CONDUCTORS

Type	Conductor Size (AWG/kcmil)	Metric Designator (Trade Size)							
		12 (3/8)	16 (1/2)	21 (3/4)	27 (1)	35 (1-1/4)	41 (1-1/2)	53 (2)
RHH†, RHW†, RHW-2†, TW, THW, THHW, THW-2	6	1	1	3	5	8	11	18
	4	1	1	1	3	6	8	13
	3	1	1	1	3	5	7	11
	2	0	1	1	2	4	6	9
	1	0	1	1	1	3	4	7
	1/0	0	0	1	1	2	3	6
	2/0	0	0	1	1	2	3	5
	3/0	0	0	1	1	1	2	4
	4/0	0	0	0	1	1	1	3
	250	0	0	0	1	1	1	3
	300	0	0	0	1	1	1	2
	350	0	0	0	0	1	1	1
	400	0	0	0	0	1	1	1
	500	0	0	0	0	1	1	1

600	0	0	0	0	1	1	1
700	0	0	0	0	0	1	1
750	0	0	0	0	0	1	1
800	0	0	0	0	0	1	1
900	0	0	0	0	0	1	1
1000	0	0	0	0	0	0	1
1250	0	0	0	0	0	0	1
1500	0	0	0	0	0	0	0
1750	0	0	0	0	0	0	0
2000	0	0	0	0	0	0	0
THHN, THWN, THWN-2							
14	8	13	22	36	63	81	133
12	5	9	16	26	46	59	97
10	3	6	10	16	29	37	61
8	1	3	6	9	16	21	35
6	1	2	4	7	12	15	25

(continued)

Table C5 Maximum Number of Conductors or Fixture Wires in Liquidtight Flexible Nonmetallic Conduit (Type LFNC-B*) (Based on Table 1) *Continued*

CONDUCTORS

Type	Conductor Size (AWG/kcmil)	Metric Designator (Trade Size)							
		12 (3/8)	16 (1/2)	21 (3/4)	27 (1)	35 (1-1/4)	41 (1-1/2)	53 (2)
THHN, THWN, THWN-2	4	1	1	2	4	7	9	15
	3	1	1	1	3	6	8	13
	2	1	1	1	3	5	7	11
	1	0	1	1	1	4	5	8
	1/0	0	0	1	1	3	4	7
	2/0	0	0	1	1	2	3	6
	3/0	0	0	1	1	1	3	5
	4/0	0	0	1	1	1	2	4
	250	0	0	0	1	1	1	3
	300	0	0	0	1	1	1	3
	350	0	0	0	1	1	1	2
	400	0	0	0	0	1	1	1
	500	0	0	0	0	1	1	1

Type	Size							
	600	0	0	0	0	1	1	1
	700	0	0	0	0	1	1	1
	750	0	0	0	0	0	1	1
	800	0	0	0	0	0	1	1
	900	0	0	0	0	0	1	1
	1000	0	0	0	0	0	0	1
FEP, FEPB, PFA, PFAH, TFE	14	7	12	21	35	61	79	129
	12	5	9	15	25	44	57	94
	10	4	6	11	18	32	41	68
	8	1	3	6	10	18	23	39
	6	1	2	4	7	13	17	27
	4	1	1	3	5	9	12	19
	3	1	1	2	4	7	10	16
	2	1	1	1	3	6	8	13
PFA, PFAH, TFE	1	0	1	1	2	4	5	9
PFA, PFAH	1/0	0	1	1	1	3	4	7

(continued)

Table C5 Maximum Number of Conductors or Fixture Wires in Liquidtight Flexible Nonmetallic Conduit (Type LFNC-B*) (Based on Table 1) *Continued*

CONDUCTORS

Type	Conductor Size (AWG/kcmil)	Metric Designator (Trade Size)							
		12 (3/8)	16 (1/2)	21 (3/4)	27 (1)	35 (1-1/4)	41 (1-1/2)	53 (2)
TFE, Z	2/0	0	1	1	1	3	4	6
	3/0	0	0	1	1	2	3	5
	4/0	0	0	1	1	1	2	4
Z	14	9	15	26	42	73	95	156
	12	6	10	18	30	52	67	111
	10	4	6	11	18	32	41	68
	8	2	4	7	11	20	26	43
	6	1	3	5	8	14	18	30
	4	1	1	3	5	9	12	20
	3	1	1	2	4	7	9	15
	2	0	1	1	3	6	7	12
	1	0	1	1	2	5	6	10

XHH, XHHW, XHHW-2, ZW							
14	5	9	15	25	44	57	93
12	4	7	12	19	33	43	71
10	3	5	9	14	25	32	53
8	1	3	5	8	14	18	29
6	1	1	3	6	10	13	22
4	1	1	2	4	7	9	16
3	1	1	1	3	6	8	13
2	1	1	1	3	5	7	11
1	0	1	1	1	4	5	8
XHH, XHHW, XHHW-2							
1/0	0	1	1	1	3	4	7
2/0	0	0	1	1	2	3	6
3/0	0	0	1	1	1	3	5
4/0	0	0	1	1	1	2	4
250	0	0	0	1	1	1	3
300	0	0	0	1	1	1	3
350	0	0	0	1	1	1	2

(continued)

Table C5 Maximum Number of Conductors or Fixture Wires in Liquidtight Flexible Nonmetallic Conduit (Type LFNC-B*) (Based on Table 1) *Continued*

		CONDUCTORS								
		Metric Designator (Trade Size)								
Type	Conductor Size (AWG/kcmil)	12 (3/8)	16 (1/2)	21 (3/4)	27 (1)	35 (1-1/4)	41 (1-1/2)	53 (2)
XHH, XHHW, XHHW-2	400	0	0	0	0	1	1	1
	500	0	0	0	0	1	1	1
	600	0	0	0	0	1	1	1
	700	0	0	0	0	1	1	1
	750	0	0	0	0	0	1	1
	800	0	0	0	0	0	1	1
	900	0	0	0	0	0	1	1
	1000	0	0	0	0	0	0	1
	1250	0	0	0	0	0	0	1
	1500	0	0	0	0	0	0	1
	1750	0	0	0	0	0	0	0
	2000	0	0	0	0	0	0	0

Table C6 Maximum Number of Conductors or Fixture Wires in Liquidtight Flexible Nonmetallic Conduit (Type LFNC-A*) (Based on Table 1)

Type	Conductor Size (AWG/kcmil)	Metric Designator (Trade Size)						
		12 (3/8)	16 (1/2)	21 (3/4)	27 (1)	35 (1-1/4)	41 (1-1/2)	53 (2)
RHH, RHW, RHW-2	14	2	4	7	11	20	27	45
	12	1	3	6	9	17	23	38
	10	1	3	5	8	13	18	30
	8	1	1	2	4	7	9	16
	6	1	1	1	3	5	7	13
	4	0	1	1	2	4	6	10
	3	0	1	1	1	4	5	8
	2	0	1	1	1	3	4	7
	1	0	0	1	1	1	3	5
	1/0	0	0	1	1	1	2	4
	2/0	0	0	1	1	1	1	4

(continued)

Table C6 Maximum Number of Conductors or Fixture Wires in Liquidtight
Flexible Nonmetallic Conduit (Type LFNC-A*) (Based on Table 1) *Continued*

Type	Conductor Size (AWG/kcmil)	CONDUCTORS								
		Metric Designator (Trade Size)								
		12 (3/8)	16 (1/2)	21 (3/4)	27 (1)	35 (1-1/4)	41 (1-1/2)	53 (2)
RHH, RHW, RHW-2	3/0	0	0	0	1	1	1	3
	4/0	0	0	0	1	1	1	3
	250	0	0	0	0	1	1	1
	300	0	0	0	0	1	1	1
	350	0	0	0	0	1	1	1
	400	0	0	0	0	1	1	1
	500	0	0	0	0	0	1	1
	600	0	0	0	0	0	1	1
	700	0	0	0	0	0	0	1
	750	0	0	0	0	0	0	1
	800	0	0	0	0	0	0	1
	900	0	0	0	0	0	0	1

Type	Size							
	1000	1	0	0	0	0	0	0
	1250	0	0	0	0	0	0	0
	1500	0	0	0	0	0	0	0
	1750	0	0	0	0	0	0	0
	2000	0	0	0	0	0	0	0
TW	14	96	58	43	24	15	9	5
	12	74	44	33	19	12	7	4
	10	55	33	24	14	9	5	3
	8	30	18	13	8	5	3	1
RHH†, RHW†, RHW-2†, THHWO, THW, THW-2	14	64	38	28	16	10	6	3
RRH†, RHW†, RHW-2†, THHW, THW	12	51	31	23	13	8	4	3
	10	40	24	18	10	6	3	1
RHH†, RHW†, RHW-2†, THHW, THW, THW-2	8	24	14	10	6	4	1	1
RHH†, RHW†, RHW-2†, TW, THW, THHW, THW-2	6	18	11	8	4	3	1	1
	4	13	8	6	3	1	1	1

(continued)

Table C6 Maximum Number of Conductors or Fixture Wires in Liquidtight Flexible Nonmetallic Conduit (Type LFNC-A*) (Based on Table 1) *Continued*

Type	Conductor Size (AWG/kcmil)	CONDUCTORS Metric Designator (Trade Size)							
		12 (3/8)	16 (1/2)	21 (3/4)	27 (1)	35 (1-1/4)	41 (1-1/2)	53 (2)
RHH†, RHW†, RHW-2†, TW, THW, THHW, THW-2	3	1	1	1	3	5	7	11
	2	1	1	1	2	4	6	10
	1	0	1	1	1	3	4	7
RHH†, RHW†, RHW-2†, TW, THW, THHW, THW-2	1/0	0	0	1	1	2	3	6
	2/0	0	0	1	1	1	3	5
	3/0	0	0	1	1	1	2	4
	4/0	0	0	0	1	1	1	3
	250	0	0	0	1	1	1	3
	300	0	0	0	1	1	1	2
	350	0	0	0	0	1	1	1
	400	0	0	0	0	1	1	1
	500	0	0	0	0	1	1	1
	600	0	0	0	0	1	1	1
	700	0	0	0	0	0	1	1

THHN, THWN, THWN-2							
750	1	1	0	0	0	0	0
800	1	1	0	0	0	0	0
900	1	0	0	0	0	0	0
1000	1	0	0	0	0	0	0
1250	1	0	0	0	0	0	0
1500	1	0	0	0	0	0	0
1750	0	0	0	0	0	0	0
2000	0	0	0	0	0	0	0
14	137	83	62	35	22	13	8
12	100	60	45	25	16	9	5
10	63	38	28	16	10	6	3
8	36	22	16	9	6	3	1
6	26	16	12	6	4	2	1
4	16	9	7	4	2	1	1
3	13	8	6	3	1	1	1
2	11	7	5	3	1	1	1
1	8	5	4	1	1	1	0
1/0	7	4	3	1	1	1	0
2/0	6	3	2	1	1	0	0

(continued)

Table C6 Maximum Number of Conductors or Fixture Wires in Liquidtight Flexible Nonmetallic Conduit (Type LFNC-A*) (Based on Table 1) *Continued*

CONDUCTORS

Type	Conductor Size (AWG/kcmil)	Metric Designator (Trade Size)							
		12 (3/8)	16 (1/2)	21 (3/4)	27 (1)	35 (1-1/4)	41 (1-1/2)	53 (2)
THHN, THWN, THWN-2	3/0	0	0	1	1	1	3	5
	4/0	0	0	1	1	1	2	4
	250	0	0	0	1	1	1	3
	300	0	0	0	1	1	1	3
	350	0	0	0	1	1	1	2
	400	0	0	0	0	1	1	1
	500	0	0	0	0	1	1	1
	600	0	0	0	0	0	1	1
	700	0	0	0	0	1	1	1
	750	0	0	0	0	0	1	1
	800	0	0	0	0	0	1	1
	900	0	0	0	0	0	0	1
	1000	0	0	0	0	0	0	1

FEP, FEPB, PFA, PFAH, TFE	14	7	12	21	34	60	80	133
	12	5	9	15	25	44	59	97
	10	4	6	11	18	31	42	70
	8	1	3	6	10	18	24	40
	6	1	2	4	7	13	17	28
	4	1	1	3	5	9	12	20
	3	1	1	2	4	7	10	16
	2	1	1	1	3	6	8	13
PFA, PFAH, TFE	1	0	1	1	2	4	5	9
PFA, PFAH, TFE, Z	1/0	0	1	1	1	3	5	8
	2/0	0	1	1	1	3	4	6
	3/0	0	0	1	1	2	3	5
	4/0	0	0	1	1	1	2	4
Z	14	9	15	25	41	72	97	161
	12	6	10	18	29	51	69	114
	10	4	6	11	18	31	42	70
	8	2	4	7	11	20	26	44
	6	1	3	5	8	14	18	31

(continued)

Table C6 Maximum Number of Conductors or Fixture Wires in Liquidtight Flexible Nonmetallic Conduit (Type LFNC-A*) (Based on Table 1) *Continued*

Type	Conductor Size (AWG/kcmil)	12 (3/8)	16 (1/2)	21 (3/4)	27 (1)	35 (1-1/4)	41 (1-1/2)	53 (2)
				Metric Designator (Trade Size)				
Z	4	1	1	3	5	9	13	21
	3	1	1	2	4	7	9	15
	2	1	1	1	3	6	8	13
	1	1	1	1	2	4	6	10
XHH, XHHW, XHHW-2, ZW	14	5	9	15	24	43	58	96
	12	4	7	12	19	33	44	74
	10	3	5	9	14	24	33	55
	8	1	3	5	8	13	18	30
	6	1	1	3	5	10	13	22
	4	1	1	2	4	7	10	16
	3	1	1	1	3	6	8	14
	2	1	1	1	3	5	7	11

XHH, XHHW, XHHW-2	8	5	4	1	1	1	1	0
1	8	5	4	1	1	1	1	0
1/0	7	4	3	1	1	1	1	0
2/0	6	3	2	1	1	1	0	0
3/0	5	3	1	1	1	1	0	0
4/0	4	2	1	1	1	1	0	0
250	3	1	1	1	0	0	0	0
300	3	1	1	1	0	0	0	0
350	2	1	1	1	0	0	0	0
400	1	1	1	0	0	0	0	0
500	1	1	1	0	0	0	0	0
600	1	1	1	0	0	0	0	0
700	1	1	1	0	0	0	0	0
750	1	1	0	0	0	0	0	0
800	1	1	0	0	0	0	0	0
900	1	1	0	0	0	0	0	0
1000	1	0	0	0	0	0	0	0
1250	1	0	0	0	0	0	0	0
1500	1	0	0	0	0	0	0	0

(continued)

Table C6 Maximum Number of Conductors or Fixture Wires in Liquidtight Flexible Nonmetallic Conduit (Type LFNC-A*) (Based on Table 1) *Continued*

		CONDUCTORS							
					Metric Designator (Trade Size)				
Type	Conductor Size (AWG/kcmil)	12 (3/8)	16 (1/2)	21 (3/4)	27 (1)	35 (1-1/4)	41 (1-1/2)	53 (2)
XHH, XHHW, XHHW-2	1750	0	0	0	0	0	0	0
	2000	0	0	0	0	0	0	0

**Table C7 Maximum Number of Conductors or Fixture Wires
in Liquidtight Flexible Metal Conduit (LFMC) (Based on Table 1)**

					CONDUCTORS							
	Conductor Size (AWG/ kcmil)				Metric Designator (Trade Size)							
Type		16 (1/2)	21 (3/4)	27 (1)	35 (1-1/4)	41 (1-1/2)	53 (2)	63 (2-1/2)	78 (3)	91 (3-1/2)	103 (4)
RHH, RHW, RHW-2	14	4	7	12	21	27	44	66	102	133	173
	12	3	6	10	17	22	36	55	84	110	144
	10	3	5	8	14	18	29	44	68	89	116
	8	1	2	4	7	9	15	23	36	46	61
	6	1	1	3	6	7	12	18	28	37	48
	4	1	1	2	4	6	9	14	22	29	38
	3	1	1	1	4	5	8	13	19	25	33
	2	1	1	1	3	4	7	11	17	22	29
	1	0	1	1	1	3	5	7	11	14	19
	1/0	0	1	1	1	2	4	6	10	13	16
	2/0	0	1	1	1	1	3	5	8	11	14

(continued)

Table C7 Maximum Number of Conductors or Fixture Wires in Liquidtight Flexible Metal Conduit (LFMC) (Based on Table 1) *Continued*

Type	Conductor Size (AWG/kcmil)	16 (1/2)	21 (3/4)	27 (1)	35 (1-1/4)	41 (1-1/2)	53 (2)	63 (2-1/2)	78 (3)	91 (3-1/2)	103 (4)
					CONDUCTORS — Metric Designator (Trade Size)						
RHH, RHW, RHW-2	3/0	0	0	1	1	1	3	4	7	9	12
	4/0	0	0	1	1	1	2	4	6	8	10
	250	0	0	0	1	1	1	3	4	6	8
	300	0	0	0	1	1	1	2	4	5	7
	350	0	0	0	1	1	1	2	3	5	6
	400	0	0	0	1	1	1	1	3	4	6
	500	0	0	0	1	1	1	1	3	4	5
	600	0	0	0	0	1	1	1	2	3	4
	700	0	0	0	0	0	1	1	1	3	3
	750	0	0	0	0	0	1	1	1	2	3
	800	0	0	0	0	0	1	1	1	2	3
	900	0	0	0	0	0	1	1	1	2	3

	1000	0	0	0	0	0	1	1	1	1	3
	1250	0	0	0	0	0	0	1	1	1	1
	1500	0	0	0	0	0	0	1	1	1	1
	1750	0	0	0	0	0	0	1	1	1	1
	2000	0	0	0	0	0	0	1	1	1	1
TW	14	9	15	25	44	57	93	140	215	280	365
	12	7	12	19	33	43	71	108	165	215	280
	10	5	9	14	25	32	53	80	123	160	209
	8	3	5	8	14	18	29	44	68	89	116
RHH*, RHW*, RHW-2*, THHW, THW, THW-2	14	6	10	16	29	38	62	93	143	186	243
RHH*, RHW*, RHW-2*, THHW, THW	12	5	8	13	23	30	50	75	115	149	195
	10	3	6	10	18	23	39	58	89	117	152
RHH*, RHW*, RHW-2*, THHW, THW-2	8	1	4	6	11	14	23	35	53	70	91

(continued)

Table C7 Maximum Number of Conductors or Fixture Wires
in Liquidtight Flexible Metal Conduit (LFMC) (Based on Table 1)

CONDUCTORS

Type	Conductor Size (AWG/ kcmil)	Metric Designator (Trade Size)											
		16 (1/2)	21 (3/4)	27 (1)	35 (1-1/4)	41 (1-1/2)	53 (2)	63 (2-1/2)	78 (3)	91 (3-1/2)	103 (4)
RHH*, RHW*,	6	1	3	5	8	11	18	27	41	53	70
RHW-2*,	4	1	1	3	6	8	13	20	30	40	52
TW, THW,	3	1	1	3	5	7	11	17	26	34	44
THHW,	2	1	1	2	4	6	9	14	22	29	38
THW-2	1	1	1	1	3	4	7	10	15	20	26
RHH* RHW*,	1/0	0	0	1	2	3	6	8	13	17	23
RHW-2*,	2/0	0	1	1	2	3	5	7	11	15	19
TW, THW,	3/0	0	0	1	1	2	4	6	9	12	16
THHW,	4/0	0	0	1	1	1	3	5	8	10	13
THW-2											
	250	0	0	0	1	1	3	4	6	8	11
	300	0	0	1	1	1	2	3	5	7	9
	350	0	0	0	1	1	1	3	5	6	8
	400	0	0	0	1	1	1	3	4	6	7
	500	0	0	0	1	1	1	2	3	5	6

Size										
600	0	0	0	1	1	1	1	3	4	5
700	0	0	0	0	1	1	1	2	3	4
750	0	0	0	0	1	1	1	2	3	4
800	0	0	0	0	1	1	1	2	3	4
900	0	0	0	0	0	1	1	1	3	3
1000	0	0	0	0	0	1	1	1	2	3
1250	0	0	0	0	0	1	1	1	1	2
1500	0	0	0	0	0	0	1	1	1	2
1750	0	0	0	0	0	0	1	1	1	1
2000	0	0	0	0	0	0	1	1	1	1

THHN, THWN, THWN-2										
14	13	22	36	63	81	133	201	308	401	523
12	9	16	26	46	59	97	146	225	292	381
10	6	10	16	29	37	61	92	141	184	240
8	3	6	9	16	21	35	53	81	106	138
6	2	4	7	12	15	25	38	59	76	100
4	1	2	4	7	9	15	23	36	47	61
3	1	1	3	6	8	13	20	30	40	52
2	1	1	3	5	7	11	17	26	33	44
1	1	1	1	4	5	8	12	19	25	32

(continued)

Table C7 Maximum Number of Conductors or Fixture Wires in Liquidtight Flexible Metal Conduit (LFMC) (Based on Table 1)

CONDUCTORS

Type	Conductor Size (AWG/kcmil)	16 (1/2)	21 (3/4)	27 (1)	35 (1-1/4)	41 (1-1/2)	53 (2)	63 (2-1/2)	78 (3)	91 (3-1/2)	103 (4)
THHN, THWN, THWN-2	1/0	1	1	1	3	4	7	10	16	21	27
	2/0	0	1	1	2	3	6	8	13	17	23
	3/0	0	1	1	1	3	5	7	11	14	19
	4/0	0	1	1	1	2	4	6	9	12	15
	250	0	0	1	1	1	3	5	7	10	12
	300	0	0	1	1	1	3	4	6	8	11
	350	0	0	1	1	1	2	3	5	7	9
	400	0	0	0	1	1	1	3	5	6	8
	500	0	0	0	1	1	1	2	4	5	7
	600	0	0	0	1	1	1	1	3	4	6
	700	0	0	0	1	1	1	1	3	4	5
	750	0	0	0	0	1	1	1	3	3	5
	800	0	0	0	0	1	1	1	2	3	4
	900	0	0	0	1	1	1	1	2	3	4

Type	1000	0	0	0	0	0	1	1	1	3	3
FEP, FEPB, PFA, PFAH, TFE	14	12	21	35	61	79	129	195	299	389	507
	12	9	15	25	44	57	94	142	218	284	370
	10	6	11	18	32	41	68	102	156	203	266
	8	3	6	10	18	23	39	58	89	117	152
	6	2	4	7	13	17	27	41	64	83	108
PFA, PFAH, TFE	4	1	3	5	9	12	19	29	44	58	75
	3	1	2	4	7	10	16	24	37	48	63
	2	1	1	3	6	8	13	20	30	40	52
PFA, PFAH, TFE, Z	1	1	1	2	4	5	9	14	21	28	36
PFA, PFAH, TFE, Z	1/0	1	1	1	3	4	7	11	18	23	30
	2/0	1	1	1	3	4	6	9	14	19	25
	3/0	0	1	1	2	3	5	8	12	16	20
	4/0	0	1	1	1	2	4	6	10	13	17
Z	14	20	26	42	73	95	156	235	360	469	611
	12	14	18	30	52	67	111	167	255	332	434
	10	8	11	18	32	41	68	102	156	203	266

(continued)

Table C7 Maximum Number of Conductors or Fixture Wires in Liquidtight Flexible Metal Conduit (LFMC) (Based on Table 1)

CONDUCTORS

Type	Conductor Size (AWG/kcmil)	16 (1/2)	21 (3/4)	27 (1)	35 (1-1/4)	41 (1-1/2)	53 (2)	63 (2-1/2)	78 (3)	91 (3-1/2)	103 (4)
Z	8	5	7	11	20	26	43	64	99	129	168
	6	4	5	8	14	18	30	45	69	90	118
	4	2	3	5	9	12	20	31	48	62	81
	3	2	2	4	7	9	15	23	35	45	59
	2	1	1	3	6	7	12	19	29	38	49
	1	1	1	2	5	6	10	15	23	30	40
XHH, XHHW, XHHW-2, ZW	14	9	15	25	44	57	93	140	215	280	365
	12	7	12	19	33	43	71	108	165	215	280
	10	5	9	14	25	32	53	80	123	160	209
	8	3	5	8	14	18	29	44	68	89	116
	6	1	3	6	10	13	22	33	50	66	86
	4	1	2	4	7	9	16	24	36	48	62

Metric Designator (Trade Size)

XHH, XHHW, XHHW-2										
3	1	1	3	6	8	13	20	31	40	52
2	1	1	3	5	7	11	17	26	34	44
1	1	1	1	4	5	8	12	19	25	33
1/0	1	1	1	3	4	7	10	16	21	28
2/0	0	1	1	2	3	6	9	13	17	23
3/0	0	1	1	1	3	5	7	11	14	19
4/0	0	1	1	1	2	4	6	9	12	16
250	0	0	1	1	1	3	5	7	10	13
300	0	0	1	1	1	2	4	6	8	11
350	0	0	1	1	1	2	3	5	7	10
400	0	0	0	1	1	1	3	5	6	8
500	0	0	0	1	1	1	2	4	5	7
600	0	0	0	1	1	1	1	3	4	6
700	0	0	0	1	1	1	1	3	4	5
750	0	0	0	0	1	1	1	3	3	5
800	0	0	0	0	1	1	1	2	3	4
900	0	0	0	0	1	1	1	2	3	4

(continued)

Table C7 Maximum Number of Conductors or Fixture Wires in Liquidtight Flexible Metal Conduit (LFMC) (Based on Table 1)

CONDUCTORS

Type	Conductor Size (AWG/kcmil)	Metric Designator (Trade Size)									
		16 (1/2)	21 (3/4)	27 (1)	35 (1-1/4)	41 (1-1/2)	53 (2)	63 (2-1/2)	78 (3)	91 (3-1/2)	103 (4)
XHH, XHHW, XHHW-2	1000	0	0	0	0	0	1	1	1	—	3
	1250	0	0	0	0	0	1	1	1	3	3
	1500	0	0	0	0	0	1	1	1	1	2
	1750	0	0	0	0	0	0	1	1	1	2
	2000	0	0	0	0	0	0	1	1	1	2

*Types RHH, RHW, and RHW-2 without outer covering.

Table C8 Maximum Number of Conductors or Fixture Wires in Rigid Metal Conduit (RMC) (Based on Table 1)

Type	Conductor Size (AWG/kcmil)	CONDUCTORS											
		Metric Designator (Trade Size)											
		16 (1/2)	21 (3/4)	27 (1)	35 (1-1/4)	41 (1-1/2)	53 (2)	63 (2-1/2)	78 (3)	91 (3-1/2)	103 (4)	129 (5)	155 (6)
RHH, RHW, RHW-2	14	4	7	12	21	28	46	66	102	136	176	276	398
	12	3	6	10	17	23	38	55	85	113	146	229	330
	10	3	5	8	14	19	31	44	68	91	118	185	267
	8	1	2	4	7	10	16	23	36	48	61	97	139
	6	1	1	3	6	8	13	18	29	38	49	77	112
	4	1	1	2	4	6	10	14	22	30	38	60	87
	3	1	1	2	4	5	9	12	19	26	34	53	76
	2	1	1	1	3	4	7	11	17	23	29	46	66
	1	0	1	1	1	3	5	7	11	15	19	30	44
	1/0	0	1	1	1	2	4	6	10	13	17	26	38
	2/0	0	1	1	1	2	4	5	8	11	14	23	33

(continued)

Table C8 Maximum Number of Conductors or Fixture Wires in Rigid Metal Conduit (RMC) (Based on Table 1) *Continued*

CONDUCTORS

Type	Conductor Size (AWG/ kcmil)	16 (1/2)	21 (3/4)	27 (1)	35 (1-1/4)	41 (1-1/2)	53 (2)	63 (2-1/2)	78 (3)	91 (3-1/2)	103 (4)	129 (5)	155 (6)
							Metric Designator (Trade Size)						
RHH, RHW,	3/0	0	0	1	1	1	3	3	7	10	12	20	28
RHW-2	4/0	0	0	1	1	1	3	4	6	8	11	17	24
	250	0	0	0	1	1	1	3	4	6	8	13	18
	300	0	0	0	1	1	1	2	4	5	7	11	16
	350	0	0	0	1	1	1	2	4	5	6	10	15
	400	0	0	0	1	1	1	1	3	4	6	9	13
	500	0	0	0	1	1	1	1	3	4	5	8	11
	600	0	0	0	0	1	1	1	2	3	4	6	9
	700	0	0	0	0	1	1	1	1	3	4	6	8
	750	0	0	0	0	0	1	1	1	3	3	5	8
	800	0	0	0	0	0	1	1	1	2	3	5	7
	900	0	0	0	0	0	1	1	1	2	3	5	7

Type	Size												
	1000	0	0	0	0	0	1	1	1	1	3	4	6
	1250	0	0	0	0	0	0	1	1	1	1	3	5
	1500	0	0	0	0	0	0	1	1	1	1	3	4
	1750	0	0	0	0	0	0	1	1	1	1	2	4
	2000	0	0	0	0	0	0	0	1	1	1	2	3
TW	14	9	15	25	44	59	98	140	216	288	370	581	839
	12	7	12	19	33	45	75	107	165	221	284	446	644
	10	5	9	14	25	34	56	80	123	164	212	332	480
	8	3	5	8	14	19	31	44	68	91	118	185	267
RHH*, RHW*, RHW-2*, THHW, THW, THW-2	14	6	10	17	29	39	65	93	143	191	246	387	558
RHH*, RHW*, RHW-2*, THHW, THW	12	5	8	13	23	32	52	75	115	154	198	311	448
	10	3	6	10	18	25	41	58	90	120	154	242	350
RHH*, RHW*, RHW-2*, THHW, THW, THW-2	8	1	4	6	11	15	24	35	54	72	92	145	209

(continued)

Table C8 Maximum Number of Conductors or Fixture Wires in Rigid Metal Conduit (RMC) (Based on Table 1) *Continued*

CONDUCTORS

Type	Conductor Size (AWG/kcmil)	Metric Designator (Trade Size)											
		16 (1/2)	21 (3/4)	27 (1)	35 (1-1/4)	41 (1-1/2)	53 (2)	63 (2-1/2)	78 (3)	91 (3-1/2)	103 (4)	129 (5)	155 (6)
RHH*, RHW*, RHW-2*, TW, THW, THHW, THW-2	6	1	3	5	8	11	18	27	41	55	71	111	160
	4	1	1	3	6	8	14	20	31	41	53	83	120
	3	1	1	3	5	7	12	17	26	35	45	71	103
	2	1	1	2	4	6	10	14	22	30	38	60	87
	1	1	1	1	3	4	7	10	15	21	27	42	61
	1/0	0	1	1	2	3	6	8	13	18	23	36	52
	2/0	0	1	1	2	3	5	7	11	15	19	31	44
	3/0	0	1	1	1	2	4	6	9	13	16	26	37
	4/0	0	0	1	1	1	3	5	8	10	14	21	31
	250	0	0	1	1	1	3	4	6	8	11	17	25
	300	0	0	1	1	1	2	3	5	7	9	15	22
	350	0	0	0	1	1	1	3	5	6	8	13	19
	400	0	0	0	1	1	1	3	4	6	7	12	17
	500	0	0	0	1	1	1	2	3	5	6	10	14

600	0	0	0	1	1	1	1	3	4	5	8	12
700	0	0	0	0	1	1	1	2	3	4	7	10
750	0	0	0	0	1	1	1	2	3	4	7	10
800	0	0	0	0	1	1	1	2	3	4	6	9
900	0	0	0	0	1	1	1	1	3	4	6	8
1000	0	0	0	0	0	1	1	1	2	3	5	8
1250	0	0	0	0	0	1	1	1	1	2	4	6
1500	0	0	0	0	0	1	1	1	1	2	3	5
1750	0	0	0	0	0	0	1	1	1	1	3	4
2000	0	0	0	0	0	0	1	1	1	1	3	4
14	13	22	36	63	85	140	200	309	412	531	833	1202
12	9	16	26	46	62	102	146	225	301	387	608	877
10	6	10	17	29	39	64	92	142	189	244	383	552
8	3	6	9	16	22	37	53	82	109	140	221	318
6	2	4	7	12	16	27	38	59	79	101	159	230

THHN, THWN, THWN-2

(continued)

Table C8 Maximum Number of Conductors or Fixture Wires in Rigid Metal Conduit (RMC) (Based on Table 1) *Continued*

	CONDUCTORS											
	Metric Designator (Trade Size)											
Conductor Size (AWG/kcmil)	16 (1/2)	21 (3/4)	27 (1)	35 (1-1/4)	41 (1-1/2)	53 (2)	63 (2-1/2)	78 (3)	91 (3-1/2)	103 (4)	129 (5)	155 (6)
Type												
THHN, THWN, THWN-2												
4	1	2	4	7	10	16	23	36	48	62	98	141
3	1	1	3	6	8	14	20	31	41	53	83	120
2	1	1	3	5	7	11	17	26	34	44	70	100
1	1	1	1	4	5	8	12	19	25	33	51	74
1/0	1	1	1	3	4	7	10	16	21	27	43	63
2/0	0	1	1	2	3	6	8	13	18	23	36	52
3/0	0	1	1	1	3	5	7	11	15	19	30	43
4/0	0	1	1	1	2	4	6	9	12	16	25	36
250	0	0	1	1	1	3	5	7	10	13	20	29
300	0	0	1	1	1	3	4	6	8	11	17	25
350	0	0	1	1	1	2	3	5	7	10	15	22
400	0	0	1	1	1	2	3	5	7	8	13	20
500	0	0	0	1	1	1	2	4	5	7	11	16

Type	Size												
FEP, FEPB, PFA, PFAH, TFE	600	0	0	0	1	1	1	1	3	4	6	9	13
	700	0	0	0	1	1	1	1	3	4	5	8	11
	750	0	0	0	1	1	1	1	3	4	5	7	11
	800	0	0	0	0	1	1	1	2	3	4	7	10
	900	0	0	0	0	1	1	1	2	3	4	6	9
	1000	0	0	0	0	1	1	1	1	3	4	6	8
	14	12	22	35	61	83	136	194	300	400	515	808	1166
	12	9	16	26	44	60	99	142	219	292	376	590	851
	10	6	11	18	32	43	71	102	157	209	269	423	610
	8	3	6	10	18	25	41	58	90	120	154	242	350
	6	2	4	7	13	17	29	41	64	85	110	172	249
PFA, PFAH, TFE	4	1	3	5	9	12	20	29	44	59	77	120	174
	3	1	2	4	7	10	17	24	37	50	64	100	145
	2	1	1	3	6	8	14	20	31	41	53	83	120
PFA, PFAH, TFE	1	1	1	2	4	6	9	14	21	28	37	57	83
PFA, PFAH, TFE, Z	1/0	1	1	1	3	5	8	11	18	24	30	48	69
	2/0	1	1	1	3	4	6	9	14	19	25	40	57

(continued)

Table C8 Maximum Number of Conductors or Fixture Wires in Rigid Metal Conduit (RMC) (Based on Table 1) *Continued*

CONDUCTORS

Type	Conductor Size (AWG/ kcmil)	16 (1/2)	21 (3/4)	27 (1)	35 (1-1/4)	41 (1-1/2)	53 (2)	63 (2-1/2)	78 (3)	91 (3-1/2)	103 (4)	129 (5)	155 (6)
PFA, PFAH, TFE, Z	3/0	0	1	1	2	3	5	8	12	16	21	33	47
	4/0	0	1	1	1	2	4	6	10	13	17	27	39
Z	14	15	26	42	73	100	164	234	361	482	621	974	1405
	12	10	18	30	52	71	116	166	256	342	440	691	997
	10	6	11	18	32	43	71	102	157	209	269	423	610
	8	4	7	11	20	27	45	64	99	132	170	267	386
	6	3	5	8	14	19	31	45	69	93	120	188	271
	4	1	3	5	9	13	22	31	48	64	82	129	186
	3	1	2	4	7	9	16	22	35	47	60	94	136
	2	1	1	3	6	8	13	19	29	39	50	78	113
	1	1	1	2	5	6	10	15	23	31	40	63	92

Metric Designator (Trade Size) columns numbered (2) through (6).

XHH, XHHW, XHHW-2, ZW												
14	9	15	25	44	59	98	140	216	288	370	581	839
12	7	12	19	33	45	75	107	165	221	284	446	644
10	5	9	14	25	34	56	80	123	164	212	332	480
8	3	5	8	14	19	31	44	68	91	118	185	267
6	1	3	6	10	14	23	33	51	68	87	137	197
4	1	2	4	7	10	16	24	37	49	63	99	143
3	1	1	3	6	8	14	20	31	41	53	84	121
2	1	1	3	5	7	12	17	26	35	45	70	101
1	1	1	1	4	5	9	12	19	26	33	52	76
XHH, XHHW, XHHW-2												
1/0	1	1	1	3	4	7	10	16	22	28	44	64
2/0	0	1	1	2	3	6	9	13	18	23	37	53
3/0	0	1	1	1	3	5	7	11	15	19	30	44
4/0	0	1	1	1	2	4	6	9	12	16	25	36
250	0	0	1	1	1	3	5	7	10	13	20	30
300	0	0	1	1	1	3	4	6	9	11	18	25
350	0	0	1	1	1	2	3	6	7	10	15	22
400	0	0	1	1	1	2	3	5	7	9	14	20
500	0	0	0	1	1	1	2	4	5	7	11	16

(continued)

Table C8 Maximum Number of Conductors or Fixture Wires in Rigid Metal Conduit (RMC) (Based on Table 1) *Continued*

CONDUCTORS

Type	Conductor Size (AWG/kcmil)	Metric Designator (Trade Size)											
		16 (1/2)	21 (3/4)	27 (1)	35 (1-1/4)	41 (1-1/2)	53 (2)	63 (2-1/2)	78 (3)	91 (3-1/2)	103 (4)	129 (5)	155 (6)
XHH, XHHW, XHHW-2	600	0	0	0	1	1	1	1	3	4	6	9	13
	700	0	0	0	1	1	1	1	3	4	5	8	11
	750	0	0	0	0	1	1	1	3	4	5	7	11
	800	0	0	0	0	1	1	1	2	3	4	7	10
	900	0	0	0	0	1	1	1	2	3	4	6	9
	1000	0	0	0	0	1	1	1	1	3	4	6	8
	1250	0	0	0	0	0	1	1	1	2	3	4	6
	1500	0	0	0	0	0	1	1	1	1	2	4	5
	1750	0	0	0	0	0	0	1	1	1	1	3	5
	2000	0	0	0	0	0	0	1	1	1	1	3	4

Table C9 Maximum Number of Conductors or Fixture Wires in Rigid PVC Conduit, Schedule 80 (Based on Table 1)

CONDUCTORS

Type	Conductor Size (AWG/kcmil)	Metric Designator (Trade Size)												
		16 (1/2)	21 (3/4)	27 (1)	35 (1-1/4)	41 (1-1/2)	53 (2)	63 (2-1/2)	78 (3)	91 (3-1/2)	103 (4)	129 (5)	155 (6)
RHH, RHW, RHW-2	14	3	5	9	17	23	39	56	88	118	153	243	349
	12	2	4	7	14	19	32	46	73	98	127	202	290
	10	1	3	6	11	15	26	37	59	79	103	163	234
	8	1	1	3	6	8	13	19	31	41	54	85	122
	6	1	1	2	4	6	11	16	24	33	43	68	98
	4	1	1	1	3	5	8	12	19	26	33	53	77
	3	0	1	1	3	4	7	11	17	23	29	47	67
	2	0	1	1	3	4	6	9	14	20	25	41	58
	1	0	1	1	1	2	4	6	9	13	17	27	38
	1/0	0	0	1	1	1	3	5	8	11	15	23	33
	2/0	0	0	1	1	1	3	4	7	10	13	20	29

(continued)

Table C9 Maximum Number of Conductors or Fixture Wires in Rigid PVC Conduit, Schedule 80 (Based on Table 1) *Continued*

CONDUCTORS

Type	Conductor Size (AWG/kcmil)	16 (1/2)	21 (3/4)	27 (1)	35 (1-1/4)	41 (1-1/2)	53 (2)	63 (2-1/2)	78 (3)	91 (3-1/2)	103 (4)	129 (5)	155 (6)
RHH, RHW, RHW-2	3/0	0	0	1	1	1	3	4	6	8	11	17	25
	4/0	0	0	0	1	1	2	3	5	7	9	15	21
	250	0	0	0	1	1	1	2	4	5	7	11	16
	300	0	0	0	1	1	1	2	3	5	6	10	14
	350	0	0	0	0	1	1	1	3	4	5	9	13
	400	0	0	0	0	1	1	1	3	4	5	8	12
	500	0	0	0	0	1	1	1	2	3	4	7	10
	600	0	0	0	0	0	1	1	1	3	3	6	8
	700	0	0	0	0	0	1	1	1	2	3	5	7
	750	0	0	0	0	0	1	1	1	2	3	5	7
	800	0	0	0	0	0	1	1	1	2	3	4	7

Metric Designator (Trade Size)

Type	Size												
	1000	0	0	0	0	0	0	1	1	1	2	4	5
	1250	0	0	0	0	0	1	1	1	1	1	3	4
	1500	0	0	0	0	0	0	1	1	1	1	2	4
	1750	0	0	0	0	0	0	0	1	1	1	2	3
	2000	0	0	0	0	0	0	0	1	1	1	1	3
TW	14	6	11	20	35	49	82	118	185	250	324	514	736
	12	5	9	15	27	38	63	91	142	192	248	394	565
	10	3	6	11	20	28	47	67	106	143	185	294	421
	8	1	3	6	11	15	26	37	59	79	103	163	234
RHH*, RHW*, RHW-2*, THHW, THW, THW-2	14	4	8	13	23	32	55	79	123	166	215	341	490
RHH*, RHW*, RHW-2*, RHHW, THW	12	3	6	10	19	26	44	63	99	133	173	274	394
	10	2	5	8	15	20	34	49	77	104	135	214	307
RHH*, RHW*, RHW-2, RHHW, THW, THW-2	8	1	3	5	9	12	20	29	46	62	81	128	184

(continued)

Table C9 Maximum Number of Conductors or Fixture Wires in Rigid PVC Conduit, Schedule 80 (Based on Table 1) *Continued*

CONDUCTORS

Type	Conductor Size (AWG/kcmil)	Metric Designator (Trade Size)											
		16 (1/2)	21 (3/4)	27 (1)	35 (1-1/4)	41 (1-1/2)	53 (2)	63 (2-1/2)	78 (3)	91 (3-1/2)	103 (4)	129 (5)	155 (6)
RHH*, RHW*, RHW-2*, TW, THW, THHW, THW-2	6	1	1	3	7	9	16	22	35	48	62	98	141
	4	1	1	3	5	7	12	17	26	35	46	73	105
	3	1	1	2	4	6	10	14	22	30	39	63	90
	2	1	1	1	3	5	8	12	19	26	33	53	77
	1	0	1	1	2	3	6	8	13	18	23	37	54
RHH*, RHW*, RHW-2*, TW, THW, THHW, THW-2	1/0	0	0	1	1	3	5	7	11	15	20	32	46
	2/0	0	0	1	1	2	4	6	10	13	17	27	39
	3/0	0	0	1	1	1	3	5	8	11	14	23	33
	4/0	0	1	1	1	1	3	4	7	9	12	19	27
	250	0	0	0	1	1	2	3	5	7	9	15	22
	300	0	0	0	1	1	1	3	5	6	8	13	19
	350	0	0	0	1	1	1	2	4	6	7	12	17

Size												
400	15	10	7	5	4	2	1	1	1	0	0	0
500	13	9	5	4	3	1	1	1	1	0	0	0
600	10	7	4	3	2	1	1	1	0	0	0	0
700	9	6	4	3	2	1	1	1	0	0	0	0
750	8	6	4	3	1	1	1	0	0	0	0	0
800	8	6	4	3	1	1	1	0	0	0	0	0
900	7	5	3	2	1	1	1	0	0	0	0	0
1000	7	5	3	2	1	1	1	0	0	0	0	0
1250	5	4	2	1	1	1	1	0	0	0	0	0
1500	4	3	1	1	1	1	0	0	0	0	0	0
1750	4	3	1	1	1	1	0	0	0	0	0	0
2000	3	2	1	1	1	0	0	0	0	0	0	0
THHN, THWN, THWN-2												
14	1055	736	464	358	265	170	118	70	51	28	17	9
12	770	537	338	261	193	124	86	51	37	20	12	6
10	485	338	213	164	122	78	54	32	23	13	7	4
8	279	195	123	95	70	45	31	18	13	7	4	2
6	202	141	89	68	51	32	22	13	9	5	3	1
4	124	86	54	42	31	20	14	8	6	3	1	1
3	105	73	46	35	26	17	12	7	5	3	1	1

(continued)

Table C9 Maximum Number of Conductors or Fixture Wires in Rigid PVC Conduit, Schedule 80 (Based on Table 1) *Continued*

CONDUCTORS

Type	Conductor Size (AWG/kcmil)	16 (1/2)	21 (3/4)	27 (1)	35 (1-1/4)	41 (1-1/2)	53 (2)	63 (2-1/2)	78 (3)	91 (3-1/2)	103 (4)	129 (5)	155 (6)
THHN, THWN, THWN-2	2	1	1	2	4	6	10	14	22	30	39	61	88
	1	0	1	1	3	4	7	10	16	22	29	45	65
	1/0	0	1	1	2	3	6	9	14	18	24	38	55
	2/0	0	1	1	1	3	5	7	11	15	20	32	46
	3/0	0	1	1	1	2	4	6	9	13	17	26	38
	4/0	0	0	1	1	1	3	5	8	10	14	22	31
	250	0	0	1	1	1	3	4	6	8	11	18	25
	300	0	0	0	1	1	2	3	5	7	9	15	22
	350	0	0	0	1	1	1	3	5	6	8	13	19
	400	0	0	0	1	1	1	3	4	6	7	12	17
	500	0	0	0	1	1	1	2	3	5	6	10	14
	600	0	0	0	0	1	1	1	3	4	5	8	12
	700	0	0	0	0	1	1	1	2	3	4	7	10

Metric Designator (Trade Size)

Type	Size												
	750	9	7	4	3	2	1	1	1	0	0	0	0
	800	9	6	4	3	2	1	1	1	0	0	0	0
	900	8	6	3	3	1	1	1	0	0	0	0	0
	1000	7	5	3	2	1	1	1	0	0	0	0	0
FEP, FEPB, PFA, PFAH, TFE	14	1024	714	450	347	257	164	115	68	49	27	16	8
	12	747	521	328	253	188	120	84	50	36	20	12	6
	10	536	374	235	182	135	86	60	36	26	14	8	4
	8	307	214	135	104	77	49	34	20	15	8	5	2
	6	218	152	96	74	55	35	24	14	10	6	3	1
	4	153	106	67	52	38	24	17	10	7	4	2	1
	3	127	89	56	43	32	20	14	8	6	3	1	1
	2	105	73	46	35	26	17	12	7	5	3	1	1
PFA, PFAH, TFE	1	73	51	32	25	18	11	8	5	3	1	1	1
PFA, PFAH, TFE, Z	1/0	61	42	27	20	15	10	7	4	3	1	1	0
	2/0	50	35	22	17	12	8	5	3	2	1	1	0
	3/0	41	29	18	14	10	6	4	2	1	1	1	0
	4/0	34	24	15	11	8	5	4	1	1	1	0	0

(continued)

Table C9 Maximum Number of Conductors or Fixture Wires
in Rigid PVC Conduit, Schedule 80 (Based on Table 1) *Continued*

CONDUCTORS

Type	Conductor Size (AWG/kcmil)	Metric Designator (Trade Size)											
		16 (1/2)	21 (3/4)	27 (1)	35 (1-1/4)	41 (1-1/2)	53 (2)	63 (2-1/2)	78 (3)	91 (3-1/2)	103 (4)	129 (5)	155 (6)
Z	14	10	19	33	59	82	138	198	310	418	542	860	1233
	12	7	14	23	42	58	98	141	220	297	385	610	875
	10	4	8	14	26	36	60	86	135	182	235	374	536
	8	3	5	9	16	22	38	54	85	115	149	236	339
	6	2	4	6	11	16	26	38	60	81	104	166	238
Z	4	1	2	4	8	11	18	26	41	55	72	114	164
	3	1	2	3	5	8	13	19	30	40	52	83	119
	2	1	1	2	5	6	11	16	25	33	43	69	99
	1	0	1	2	4	5	9	13	20	27	35	56	80
XHH, XHHW, XHHW-2, ZW	14	6	11	20	35	49	82	118	185	250	324	514	736
	12	5	9	15	27	38	63	91	142	192	248	394	565
	10	3	6	11	20	28	47	67	106	143	185	294	421
	8	1	3	6	11	15	26	37	59	79	103	163	234

6	173	121	76	59	43	28	19	11	8	4	2	1
4	125	87	55	42	31	20	14	8	6	3	1	1
3	106	74	47	36	26	17	12	7	5	3	1	1
2	89	62	39	30	22	14	10	6	4	2	1	1
1	66	46	29	22	16	10	7	4	3	1	1	0
1/0	56	39	24	19	14	9	6	3	2	1	1	0
2/0	46	32	20	16	11	7	5	3	1	1	1	0
3/0	38	27	17	13	9	6	4	2	1	1	0	0
4/0	32	22	14	11	8	5	3	1	1	1	0	0
250	26	18	11	9	6	4	3	1	1	1	0	0
300	22	15	10	7	5	3	2	1	1	1	0	0
350	20	14	8	6	5	3	1	1	1	0	0	0
400	17	12	7	6	4	3	1	1	1	0	0	0
500	14	10	6	5	3	2	1	1	1	0	0	0
600	11	8	5	4	3	1	1	1	0	0	0	0
700	10	7	4	3	2	1	1	1	0	0	0	0
750	9	6	4	3	2	1	1	1	0	0	0	0

XHH, XHHW, XHHW-2

(continued)

Table C9 Maximum Number of Conductors or Fixture Wires in Rigid PVC Conduit, Schedule 80 (Based on Table 1) *Continued*

CONDUCTORS

Type	Conductor Size (AWG/kcmil)	Metric Designator (Trade Size)											
		16 (1/2)	21 (3/4)	27 (1)	35 (1-1/4)	41 (1-1/2)	53 (2)	63 (2-1/2)	78 (3)	91 (3-1/2)	103 (4)	129 (5)	155 (6)
XHH, XHHW, XHHW-2	800	0	0	0	0	1	1	1	1	3	4	6	9
	900	0	0	0	0	0	1	1	—	3	3	5	8
	1000	0	0	0	0	0	1	1	1	2	3	5	7
	1250	0	0	0	0	0	0	1	1	1	2	4	6
	1500	0	0	0	0	0	0	1	1	1	1	3	5
	1750	0	0	0	0	0	0	1	1	1	1	3	4
	2000	0	0	0	0	0	0	1	1	1	1	2	4

Table C10 Maximum Number of Conductors or Fixture Wires in Rigid PVC Conduit, Schedule 40 and HDPE Conduit (Based on Table 1)

						CONDUCTORS							
						Metric Designator (Trade Size)							
Type	Conductor Size (AWG/kcmil)	16 (1/2)	21 (3/4)	27 (1)	35 (1-1/4)	41 (1-1/2)	53 (2)	63 (2-1/2)	78 (3)	91 (3-1/2)	103 (4)	129 (5)	155 (6)
RHH, RHW, RHW-2	14	4	7	11	20	27	45	64	99	133	171	269	390
	12	3	5	9	16	22	37	53	82	110	142	224	323
	10	2	4	7	13	18	30	43	66	89	115	181	261
	8	1	2	4	7	9	15	22	35	46	60	94	137
	6	1	1	3	5	7	12	18	28	37	48	76	109
	4	1	1	2	4	6	10	14	22	29	37	59	85
	3	1	1	1	3	5	8	12	19	25	33	52	75
	2	1	1	1	3	4	7	10	16	22	28	45	65
	1	0	1	1	1	3	5	7	11	14	19	29	43
	1/0	0	0	1	1	2	4	6	9	13	16	26	37
	2/0	0	0	1	1	1	3	5	8	11	14	22	32

(continued)

Table C10 Maximum Number of Conductors or Fixture Wires in Rigid PVC Conduit, Schedule 40 and HDPE Conduit (Based on Table 1) *Continued*

CONDUCTORS

Type	Conductor Size (AWG/kcmil)	Metric Designator (Trade Size)											
		16 (1/2)	21 (3/4)	27 (1)	35 (1-1/4)	41 (1-1/2)	53 (2)	63 (2-1/2)	78 (3)	91 (3-1/2)	103 (4)	129 (5)	155 (6)
RHH, RHW, RHW-2	3/0	0	0	1	1	1	3	4	7	9	12	19	28
	4/0	0	0	1	1	1	2	4	6	8	10	16	24
	250	0	0	0	1	1	1	3	4	6	8	12	18
	300	0	0	0	1	1	1	2	4	5	7	11	16
	350	0	0	0	1	1	1	2	3	5	6	10	14
	400	0	0	0	1	1	1	1	3	4	6	9	13
	500	0	0	0	0	1	1	1	3	4	5	8	11
	600	0	0	0	0	1	1	1	2	3	4	6	9
	700	0	0	0	0	0	1	1	1	3	3	6	8
	750	0	0	0	0	0	1	1	1	2	3	5	8
	800	0	0	0	0	0	1	1	1	2	3	5	7
	900	0	0	0	0	0	1	1	1	2	3	5	7

Type	Size												
	1000	0	0	0	0	0	1	1	1	3	4	6	
	1250	0	0	0	0	0	0	1	1	1	3	5	
	1500	0	0	0	0	0	0	1	1	1	3	4	
	1750	0	0	0	0	0	0	1	1	1	2	3	
	2000	0	0	0	0	0	0	0	1	1	2	3	
TW	14	8	14	24	42	57	94	135	209	280	361	568	822
	12	6	11	18	32	44	72	103	160	215	277	436	631
	10	4	8	13	24	32	54	77	119	160	206	325	470
	8	2	4	7	13	18	30	43	66	89	115	181	261
RHH*, RHW*, RHW-2*, THHW, THW, THW-2	14	5	9	16	28	38	63	90	139	186	240	378	546
RHH*, RHW*, RHW-2*, THHW, THW	12	4	8	12	22	30	50	72	112	150	193	304	439
	10	3	6	10	17	24	39	56	87	117	150	237	343
RHH*, RHW*, RHW-2*, THHW, THW, THW-2	8	1	3	6	10	14	23	33	52	70	90	142	205

(continued)

Table C10 Maximum Number of Conductors or Fixture Wires in Rigid PVC Conduit, Schedule 40 and HDPE Conduit (Based on Table 1) *Continued*

CONDUCTORS

Type	Conductor Size (AWG/kcmil)	16 (1/2)	21 (3/4)	27 (1)	35 (1-1/4)	41 (1-1/2)	53 (2)	63 (2-1/2)	78 (3)	91 (3-1/2)	103 (4)	129 (5)	155 (6)
RHH*, RHW*, RHW-2*, TW, THW, THHW, THW-2	6	1	2	4	8	11	18	26	40	53	69	109	157
	4	1	1	3	6	8	13	19	30	40	51	81	117
	3	1	1	3	5	7	11	16	25	34	44	69	100
	2	1	1	2	4	6	10	14	22	29	37	59	85
	1	0	1	1	3	4	7	10	15	20	26	41	60
RHH*, RHW*, RHW-2*, TW, THW, THHW, THW-2	1/0	0	1	1	2	3	6	8	13	17	22	35	51
	2/0	0	1	1	1	3	5	7	11	15	19	30	43
	3/0	0	1	1	1	2	4	6	9	12	16	25	36
	4/0	0	0	1	1	1	3	5	8	10	13	21	30
	250	0	0	1	1	1	3	4	6	8	11	17	25
	300	0	0	1	1	2	3	5	7	9	15	21
	350	0	0	0	1	1	1	3	5	6	8	13	19

(continued)

Size												
400	17	12	7	6	4	3	1	1	1	0	0	0
500	14	10	6	5	3	2	1	1	1	0	0	0
600	11	8	5	4	3	1	1	1	0	0	0	0
700	10	7	4	3	2	1	1	1	0	0	0	0
750	9	6	4	3	2	1	1	1	0	0	0	0
800	8	6	4	3	2	1	1	1	0	0	0	0
900	8	6	3	3	1	1	1	0	0	0	0	0
1000	7	5	3	2	1	1	1	0	0	0	0	0
1250	6	4	2	1	1	1	1	0	0	0	0	0
1500	5	3	1	1	1	1	1	0	0	0	0	0
1750	4	3	1	1	1	1	0	0	0	0	0	0
2000	4	3	1	1	1	1	0	0	0	0	0	0
THHN, THWN, THWN-2												
14	1178	815	517	401	299	193	135	82	60	34	21	11
12	859	594	377	293	218	141	99	59	43	25	15	8
10	541	374	238	184	137	89	62	37	27	15	9	5
8	312	216	137	106	79	51	36	21	16	9	5	3
6	225	156	99	77	57	37	26	15	11	6	4	1
4	138	96	61	47	35	22	16	9	7	4	2	1
3	117	81	51	40	30	19	13	8	6	3	1	1

Table C10 Maximum Number of Conductors or Fixture Wires in Rigid PVC Conduit, Schedule 40 and HDPE Conduit (Based on Table 1) *Continued*

CONDUCTORS

Type	Conductor Size (AWG/kcmil)	Metric Designator (Trade Size)											
		16 (1/2)	21 (3/4)	27 (1)	35 (1-1/4)	41 (1-1/2)	53 (2)	63 (2-1/2)	78 (3)	91 (3-1/2)	103 (4)	129 (5)	155 (6)
THHN, THWN, THWN-2	2	1	1	3	5	7	11	16	25	33	43	68	98
	1	1	1	1	3	5	8	12	18	25	32	50	73
	1/0	1	1	1	3	4	7	10	15	21	27	42	61
	2/0	0	1	1	2	3	6	8	13	17	22	35	51
	3/0	0	1	1	1	3	5	7	11	14	18	29	42
	4/0	0	1	1	1	2	4	6	9	12	15	24	35
	250	0	0	1	1	1	3	4	7	10	12	20	28
	300	0	0	1	1	1	3	4	6	8	11	17	24
	350	0	0	1	1	1	2	3	5	7	9	15	21
	400	0	0	0	1	1	2	3	5	6	8	13	19
	500	0	0	0	1	1	1	2	4	5	7	11	16
	600	0	0	0	1	1	1	1	3	4	5	9	13
	700	0	0	0	0	1	1	1	3	4	5	8	11

Type	Size												
	750	0	0	0	0	1	1	1	2	3	4	7	11
	800	0	0	0	0	1	1	1	2	3	4	7	10
	900	0	0	0	0	1	1	1	2	3	4	6	9
	1000	0	0	0	0	0	1	1	1	3	3	6	8
FEP, FEPB, PFA, PFAH, TFE	14	11	20	33	58	79	131	188	290	389	502	790	1142
	12	8	15	24	42	58	96	137	212	284	366	577	834
	10	6	10	17	30	41	69	98	152	204	263	414	598
	8	3	6	10	17	24	39	56	87	117	150	237	343
	6	2	4	7	12	17	28	40	62	83	107	169	244
	4	1	3	5	8	12	19	28	43	58	75	118	170
	3	1	2	4	7	10	16	23	36	48	62	98	142
	2	1	1	3	6	8	13	19	30	40	51	81	117
PFA, PFAH, TFE	1	1	1	2	4	5	9	13	20	28	36	56	81
PFA, PFAH, TFE, Z	1/0	1	1	1	3	4	8	11	17	23	30	47	68
	2/0	0	1	1	3	4	6	9	14	19	24	39	56
	3/0	0	1	1	2	3	5	7	12	16	20	32	46
	4/0	0	1	1	1	2	4	6	9	13	16	26	38

(continued)

Table C10 Maximum Number of Conductors or Fixture Wires in Rigid PVC Conduit, Schedule 40 and HDPE Conduit (Based on Table 1) *Continued*

CONDUCTORS

Type	Conductor Size (AWG/kcmil)	Metric Designator (Trade Size)											
		16 (1/2)	21 (3/4)	27 (1)	35 (1-1/4)	41 (1-1/2)	53 (2)	63 (2-1/2)	78 (3)	91 (3-1/2)	103 (4)	129 (5)	155 (6)
Z	14	13	24	40	70	95	158	226	350	469	605	952	1376
	12	9	17	28	49	68	112	160	248	333	429	675	976
	10	6	10	17	30	41	69	98	152	204	263	414	598
	8	3	6	11	19	26	43	62	96	129	166	261	378
Z	6	2	4	7	13	18	30	43	67	90	116	184	265
	4	1	3	5	9	12	21	30	46	62	80	126	183
	3	1	2	4	6	9	15	22	34	45	58	92	133
	2	1	1	3	5	7	12	18	28	38	49	77	111
	1	1	1	2	4	6	10	14	23	30	39	62	90
XHH, XHHW, XHHW-2, ZW	14	8	14	24	42	57	94	135	209	280	361	568	822
	12	6	11	18	32	44	72	103	160	215	277	436	631
	10	4	8	13	24	32	54	77	119	160	206	325	470
	8	2	4	7	13	18	30	43	66	89	115	181	261

Size												
6	1	3	5	10	13	22	32	49	66	85	134	193
4	1	2	4	7	9	16	23	35	48	61	97	140
3	1	1	3	6	8	13	19	30	40	52	82	118
2	1	1	3	5	7	11	16	25	34	44	69	99
1	1	1	1	3	5	8	12	19	25	32	51	74
1/0	0	1	1	3	4	7	10	16	21	27	43	62
2/0	0	1	1	2	3	6	8	13	17	23	36	52
3/0	0	1	1	1	3	5	7	11	14	19	30	43
4/0	0	1	1	1	2	4	6	9	12	15	24	35
250	0	0	1	1	1	3	5	7	10	13	20	29
300	0	0	1	1	1	3	4	6	8	11	17	25
350	0	0	1	1	1	2	3	5	7	9	15	22
400	0	0	0	1	1	1	3	5	6	8	13	19
500	0	0	0	1	1	1	2	4	5	7	11	16
600	0	0	0	1	1	1	1	3	4	5	9	13
700	0	0	0	0	1	1	1	3	4	5	8	11
750	0	0	0	0	1	1	1	2	3	4	7	11

XHH, XHHW, XHHW-2

(continued)

Table C10 Maximum Number of Conductors or Fixture Wires in Rigid PVC Conduit, Schedule 40 and HDPE Conduit (Based on Table 1) *Continued*

CONDUCTORS

Type	Conductor Size (AWG/kcmil)	Metric Designator (Trade Size)												
		16 (1/2)	21 (3/4)	27 (1)	35 (1-1/4)	41 (1-1/2)	53 (2)	63 (2-1/2)	78 (3)	91 (3-1/2)	103 (4)	129 (5)	155 (6)
XHH, XHHW, XHHW-2	800	0	0	0	0	1	1	1	2	3	4	7	10
	900	0	0	0	0	1	1	1	2	3	4	6	9
	1000	0	0	0	0	0	1	1	1	3	3	6	8
	1250	0	0	0	0	0	1	1	1	1	3	4	6
	1500	0	0	0	0	0	1	1	1	1	2	4	5
	1750	0	0	0	0	0	0	1	1	1	1	3	5
	2000	0	0	0	0	0	0	1	1	1	1	3	4

INDEX